Copyright @2021 Kingschool Edition

All Right Reserved

If you have any suggestions on how to improve our book, what we can change or add to make them more useful particulary to your children please don't hesitate to contact us at kamal.elhattab@gmail.com we would be happy to hear from you.

Thank you for your trust by choosing our books and please support us leaving a review

Geometry !

Table of contents

Perimeter Rectangles

Perimeter Triangles

Perimeter Parallelograms

Area Rectangles

Area Triangles

Area Parallelograms

Volume Rectangular Prisms

Volume Triangular Prisms

Volume Cylinders

Volume Cônes

Volume Sphères

Find the perimeter.

1. 11 cm, 11 cm

2. 7 cm, 8 cm

3. 5 cm, 5 cm

4. 10 cm, 9 cm

5. 5 cm, 8 cm

6. 8 cm, 7 cm

7. 4 cm, 7 cm

8. 9 cm, 6 cm

9. 4 cm, 6 cm

10. 6 cm, 8 cm

11. 14 cm, 9 cm

12. 5 cm, 6 cm

Find the perimeter.

1.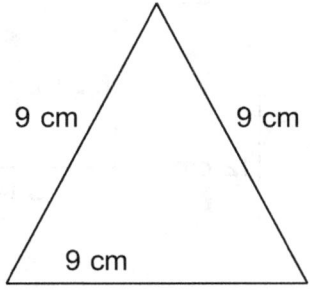

2.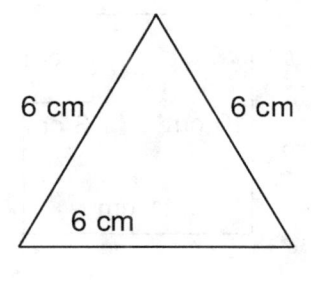

3.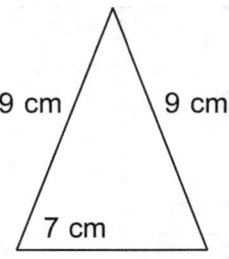

4.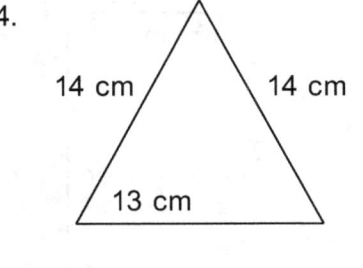

5.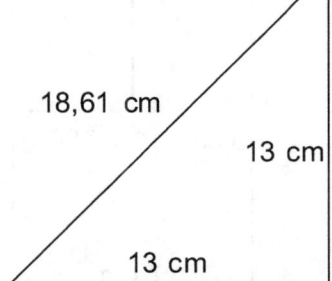

6.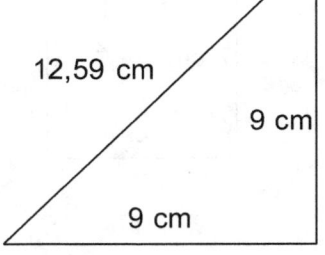

7.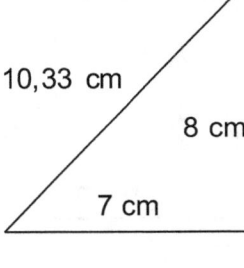

8.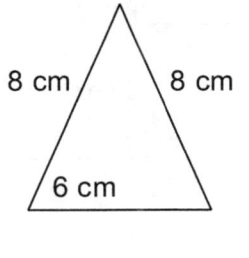

9.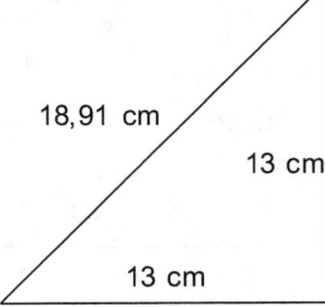

10.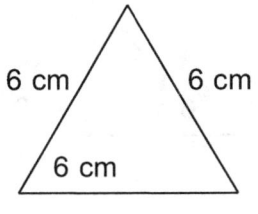

11.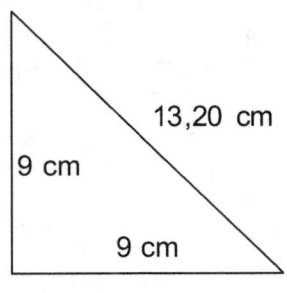

12.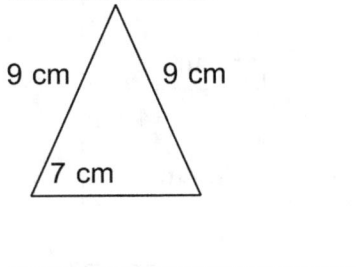

Find the perimeter.

1.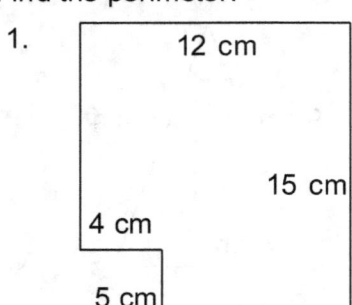

2.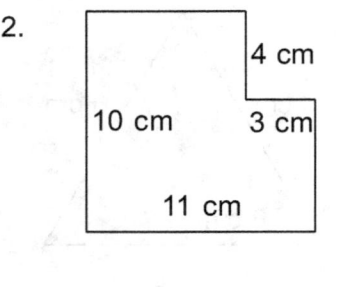

3.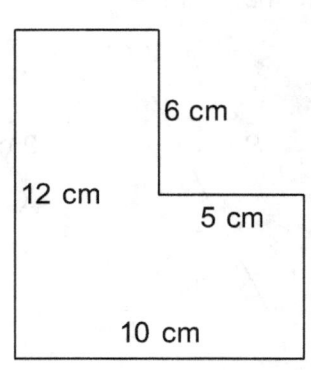

4.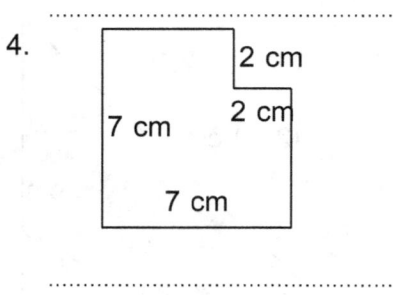

5.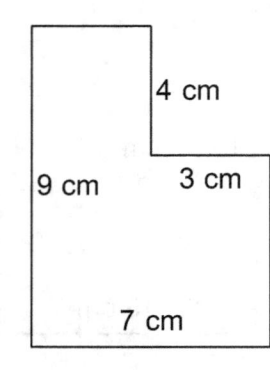

6.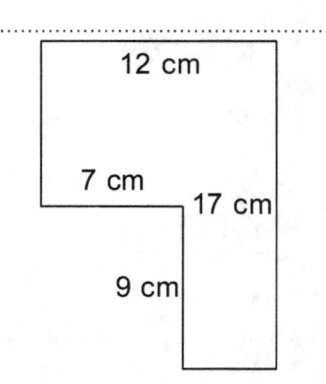

7.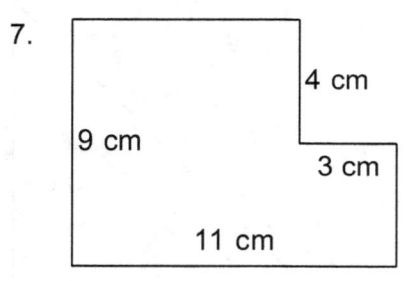

8.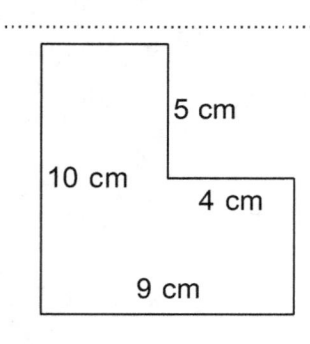

9.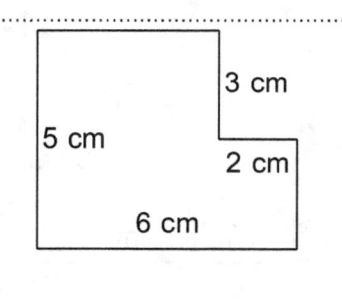

10.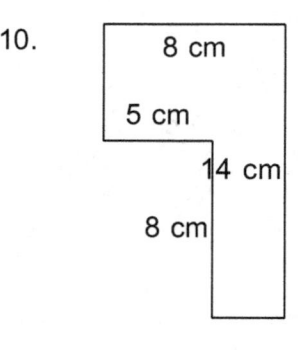

11.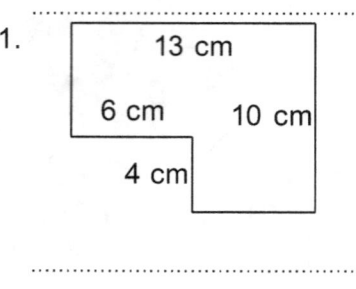

12.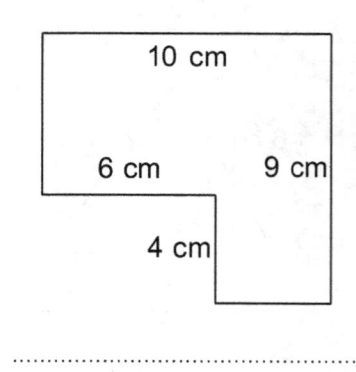

Find the perimeter.

1.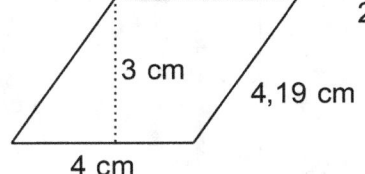

2.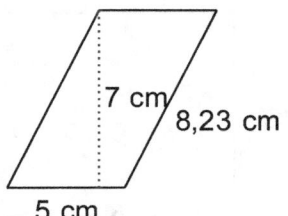

3.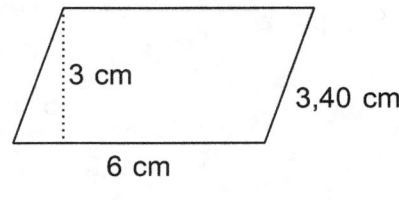

....

4.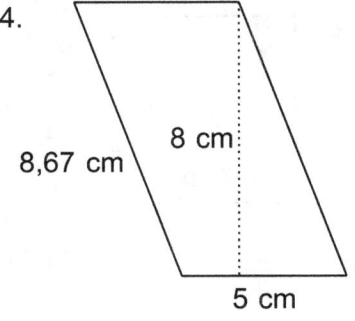

5.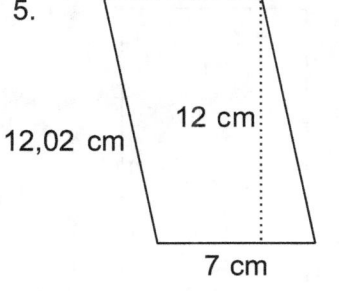

6.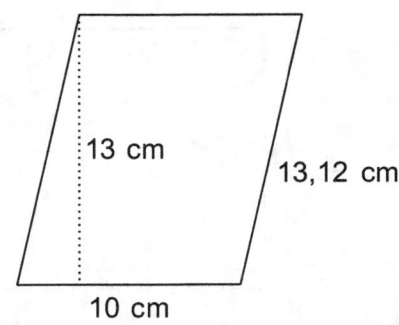

....

7.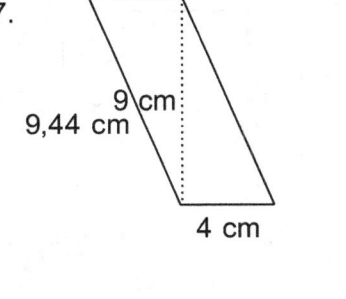

8.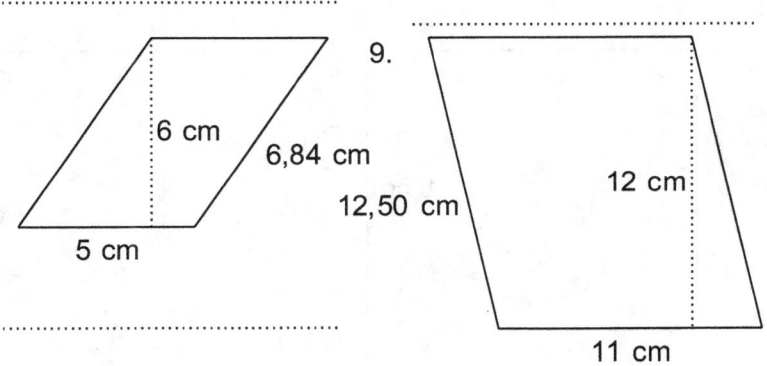

9.

....

10.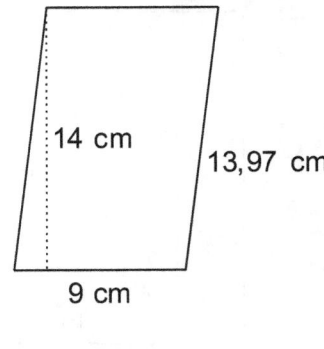

11.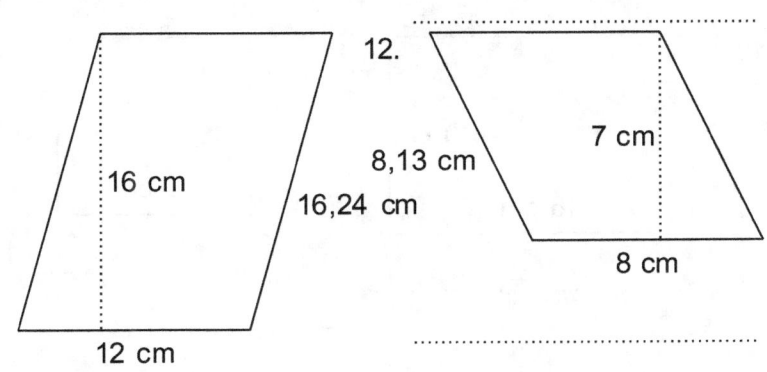

12.

....

Find the area.

1. 5 cm, 6 cm
2. 7 cm, 8 cm
3. 6 cm, 8 cm
4. 11 cm, 12 cm
5. 8 cm, 4 cm
6. 8 cm, 7 cm
7. 8 cm, 10 cm
8. 12 cm, 11 cm
9. 5 cm, 7 cm
10. 10 cm, 16 cm
11. 9 cm, 10 cm
12. 18 cm, 15 cm

Find the area.

1.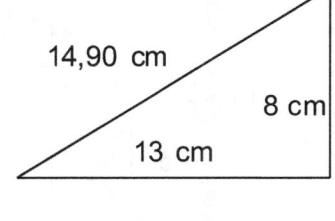
....

2.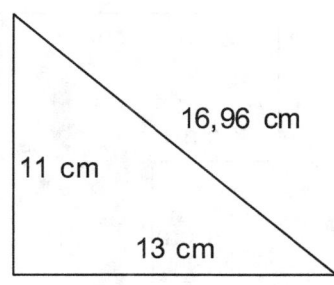
....

3.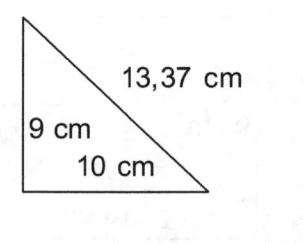
....

4.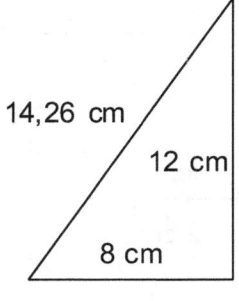
....

5.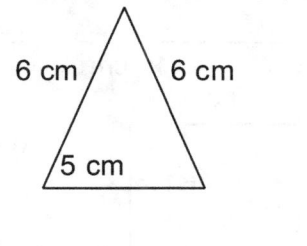
....

6.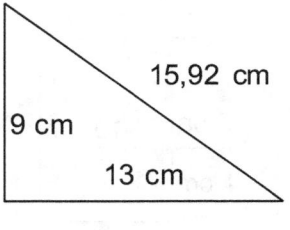
....

7.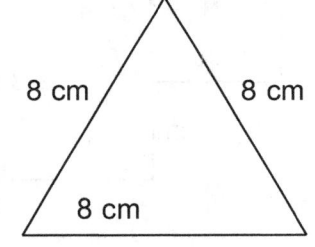
....

8.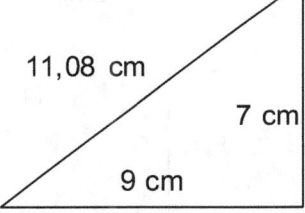
....

9.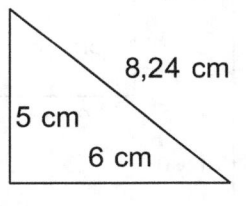
....

10.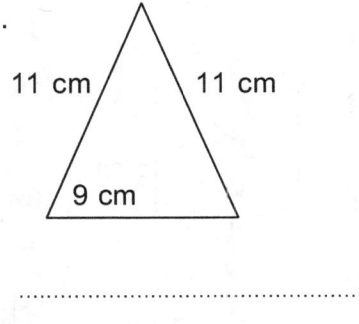
....

11.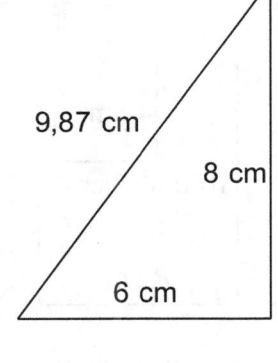
....

12.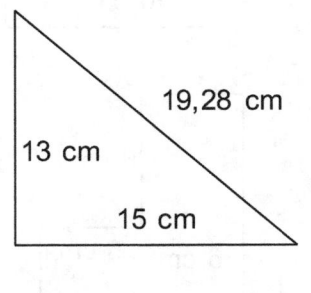
....

Find the area.

1.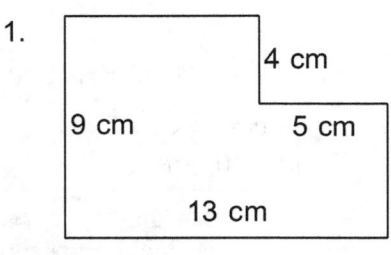

2.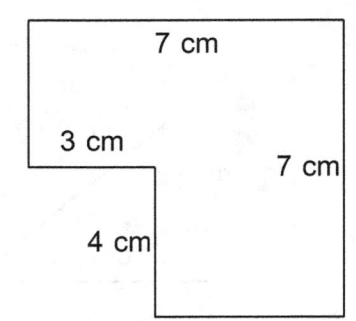

3.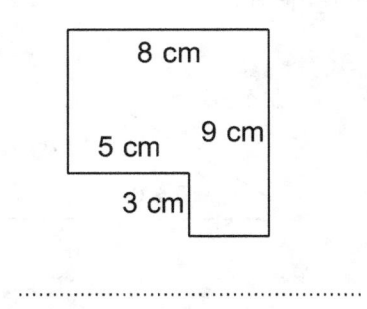

4.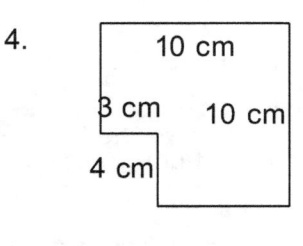

5.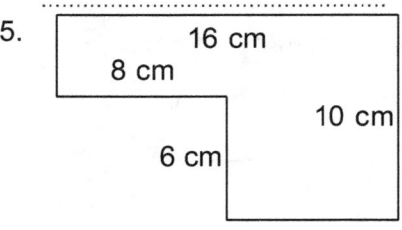

6.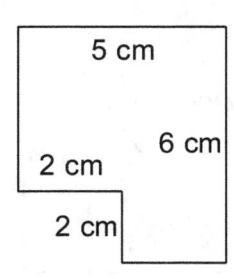

7.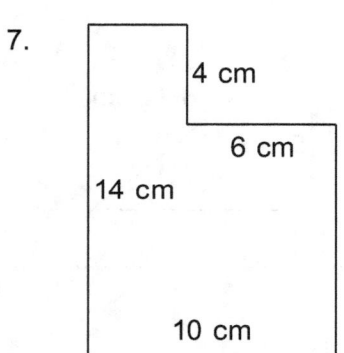

8.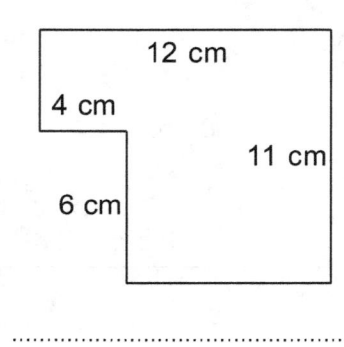

9.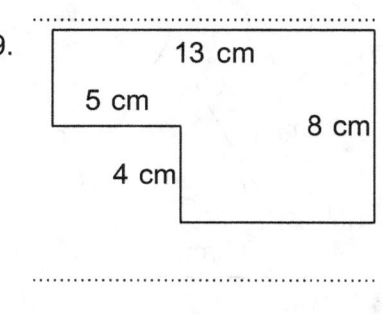

10.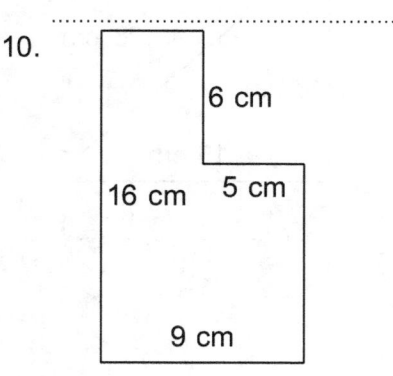

11.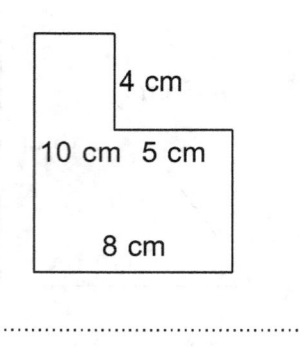

12.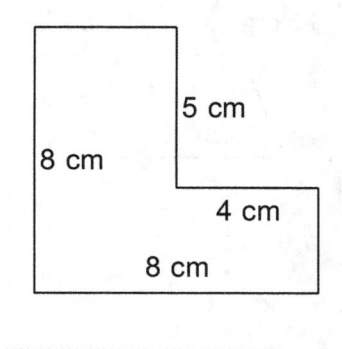

Find the area.

1.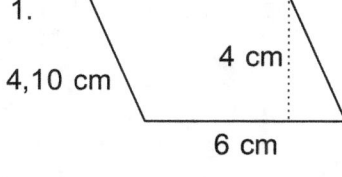

2.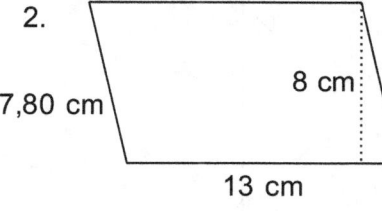

3.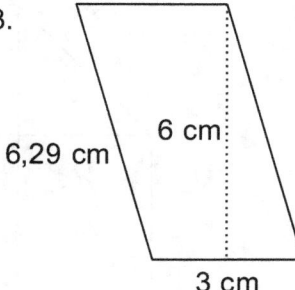

4.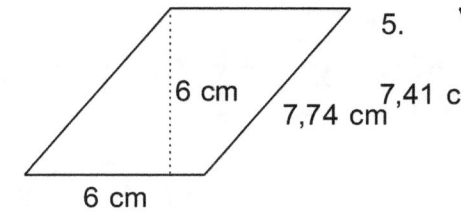

5.

6.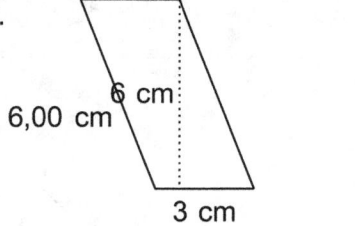

7.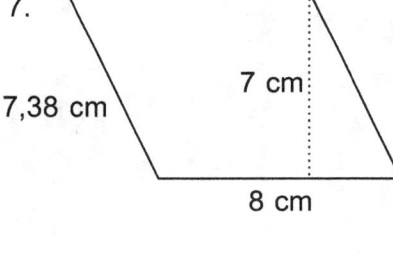

8.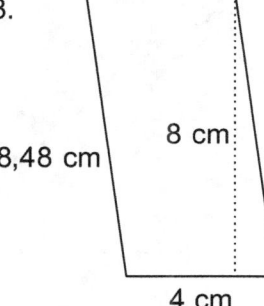

9.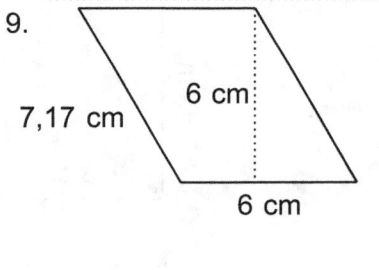

10.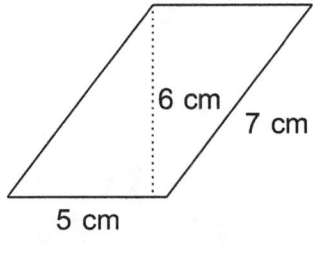

11.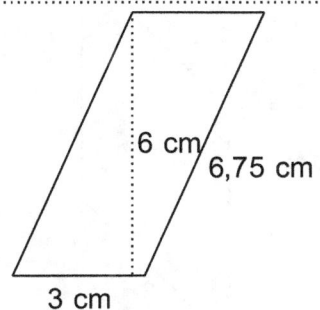

12.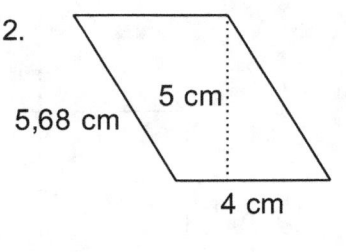

Find the volume.

1.

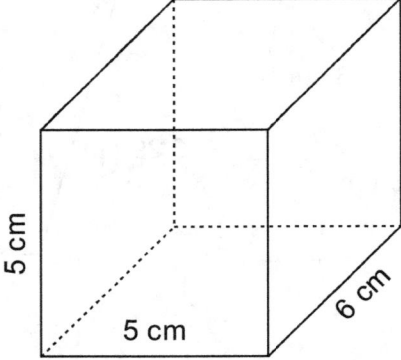

2.

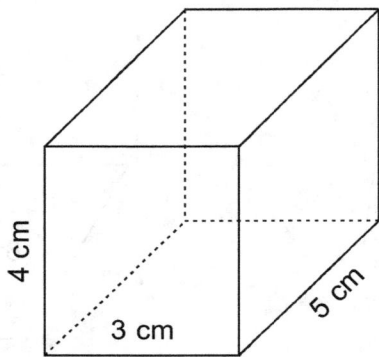

3.

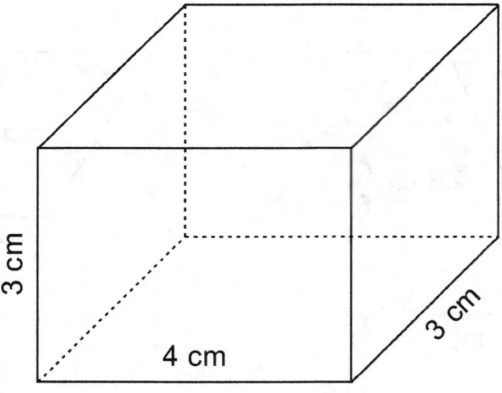

Find the volume.

1.

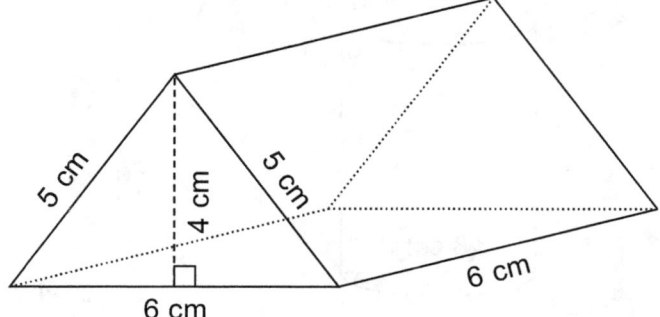

2.

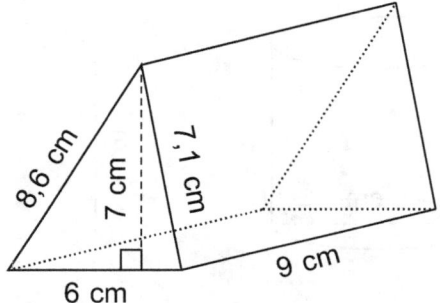

3.

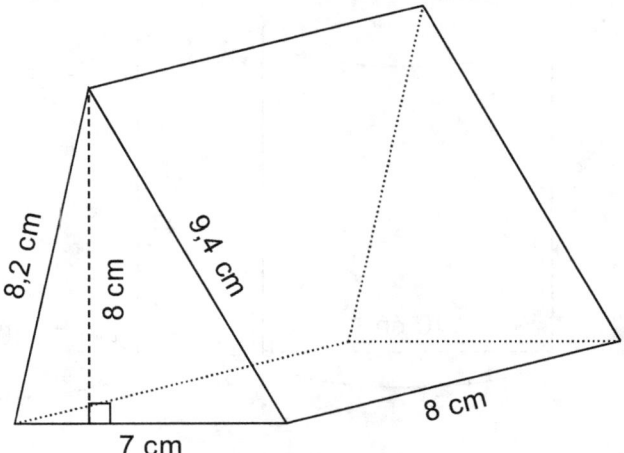

Find the volume.

1.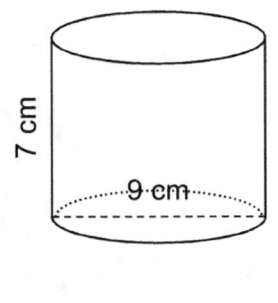

2.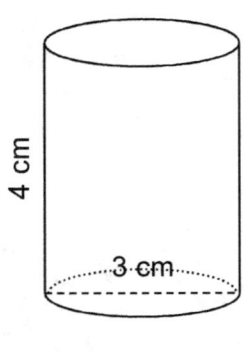

3.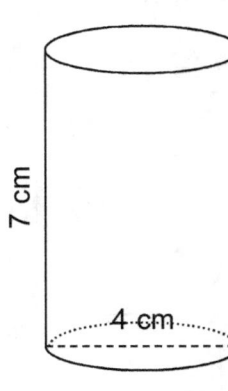

4.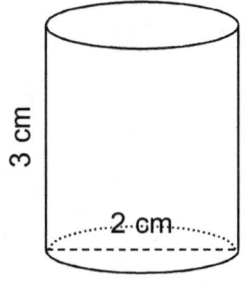

5.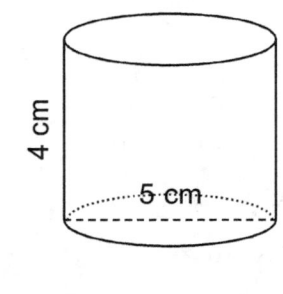

6.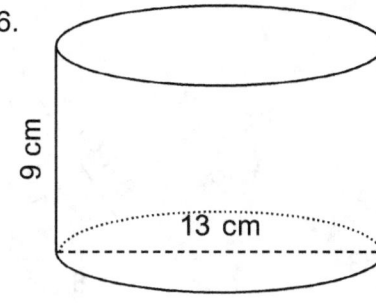

7.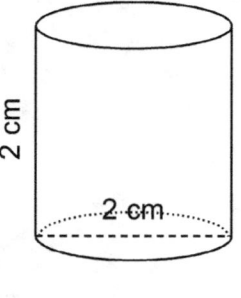

8.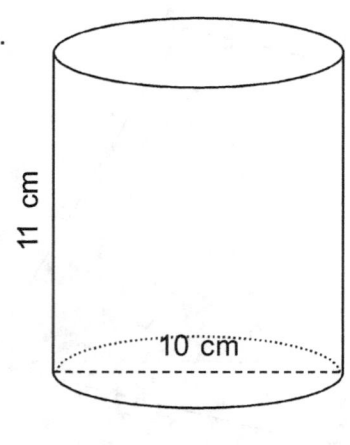

9.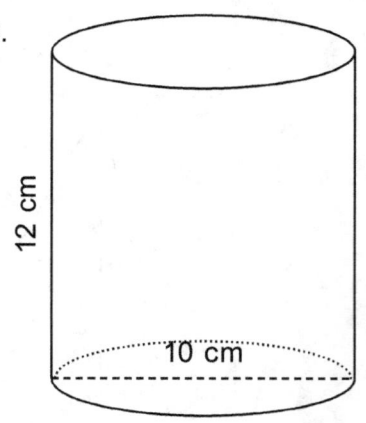

Find the volume.

1.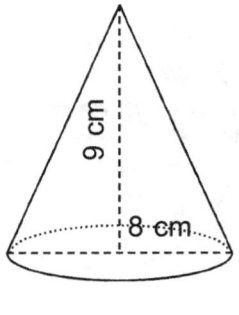
....

2.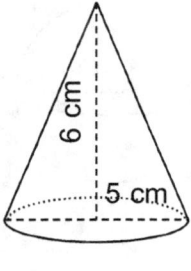
....

3.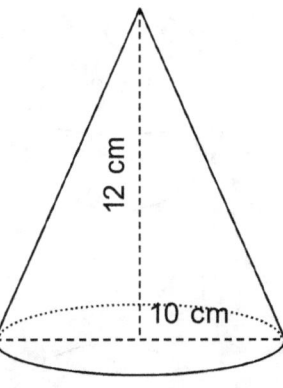
....

4.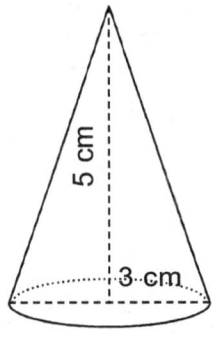
....

5.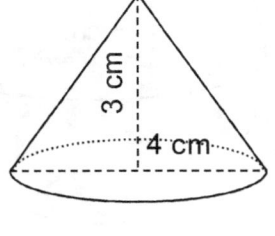
....

6.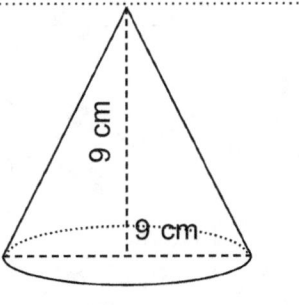
....

7.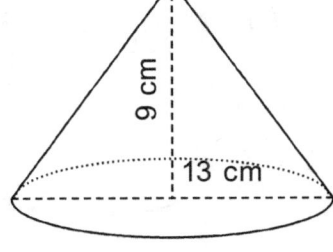
....

8.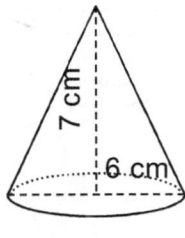
....

9.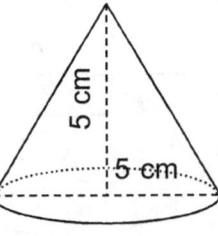
....

10.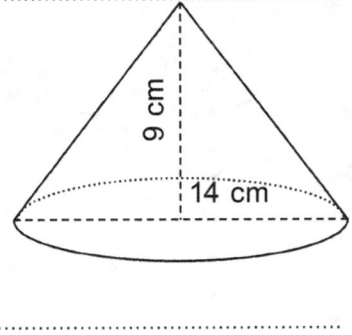
....

11.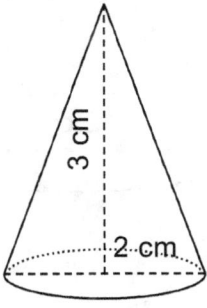
....

12.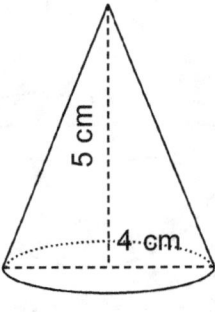
....

Find the volume.

1. 5 cm
2. 4 cm
3. 9 cm
4. 3 cm
5. 8 cm
6. 11 cm
7. 14 cm
8. 2 cm
9. 13 cm
10. 7 cm
11. 12 cm
12. 6 cm

Find the perimeter.

1. 13 cm, 10 cm

2. 5 cm, 5 cm

3. 5 cm, 5 cm

4. 16 cm, 15 cm

5. 14 cm, 11 cm

6. 9 cm, 8 cm

7. 7 cm, 11 cm

8. 12 cm, 8 cm

9. 13 cm, 12 cm

10. 5 cm, 5 cm

11. 11 cm, 11 cm

12. 7 cm, 8 cm

© KingSchool Edition

Find the perimeter.

1.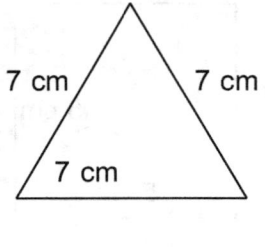
  ....

2.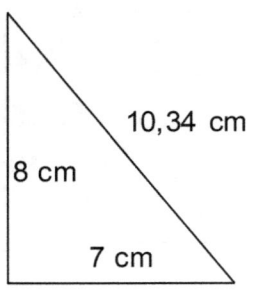
  ....

3.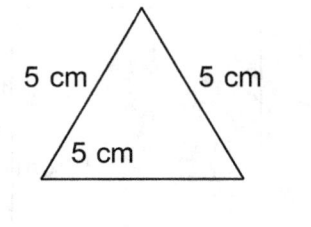
  ....

4.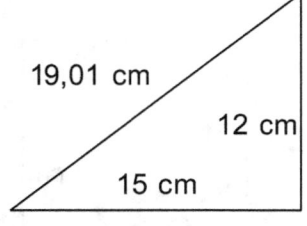
  ....

5.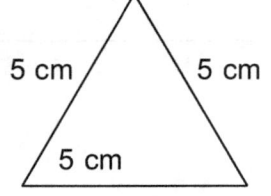
  ....

6.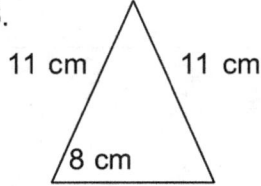
  ....

7.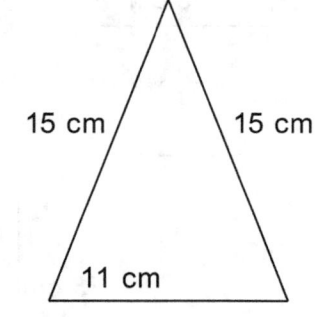
  ....

8.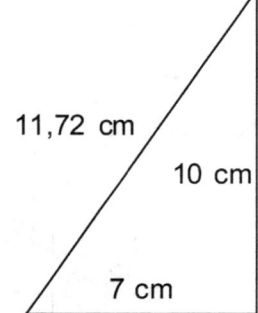
  ....

9.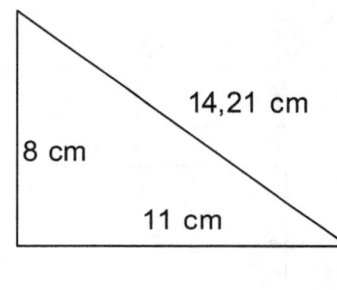
  ....

10.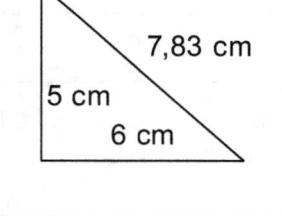
  ....

11.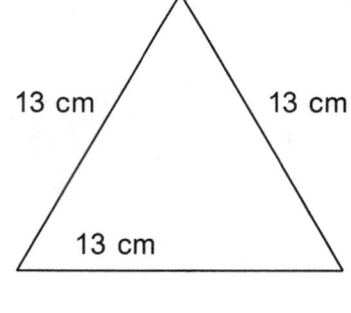
  ....

12.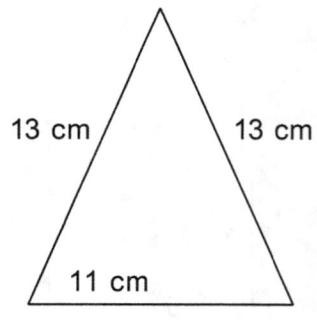
  ....

Find the perimeter.

1.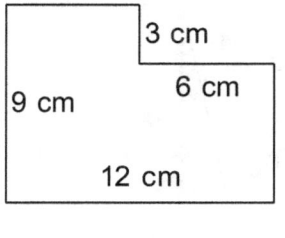

2.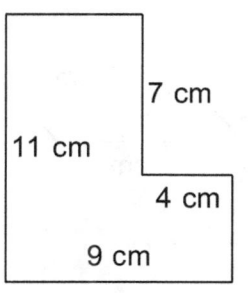

3.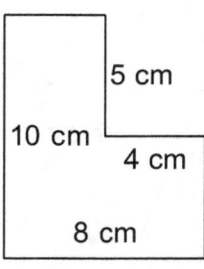

4.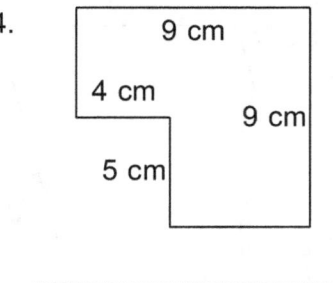

5.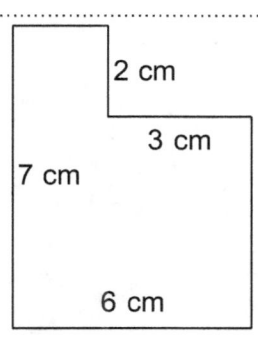

6.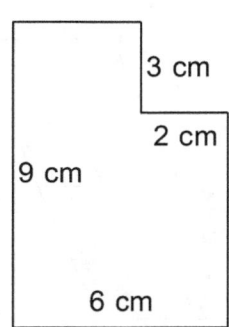

7.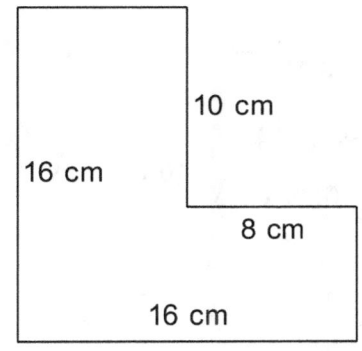

8.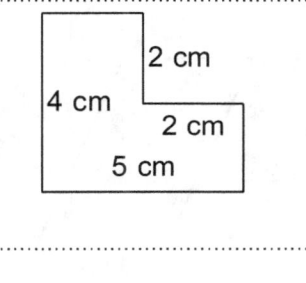

9.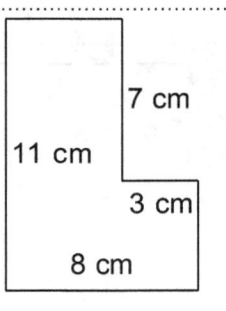

10.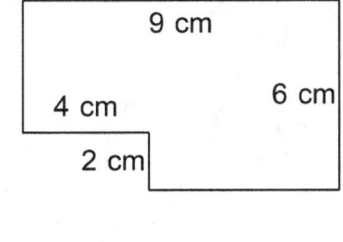

11.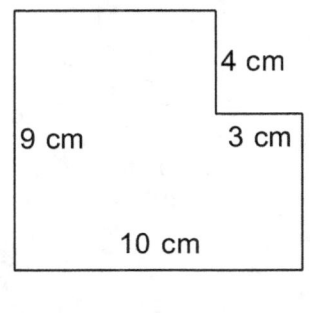

12.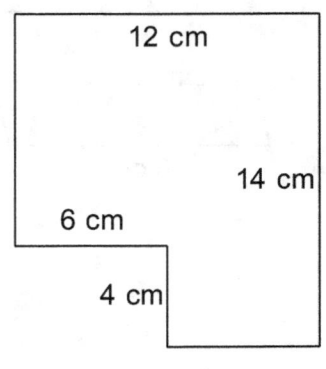

Find the perimeter.

1.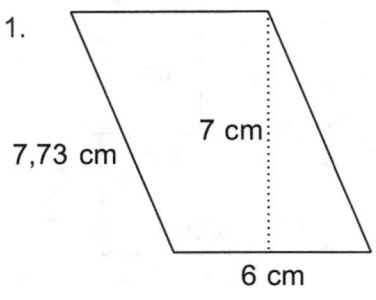

2.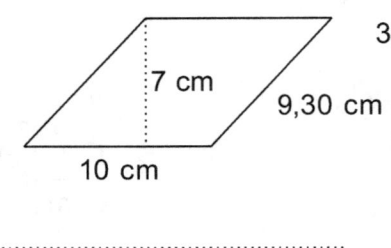

3.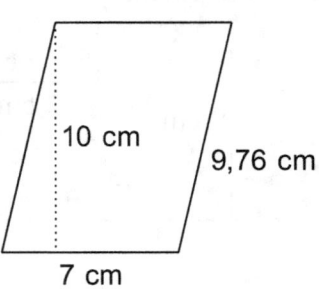

4.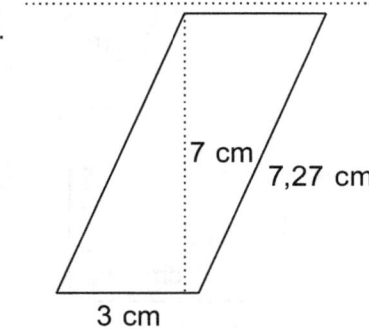

5.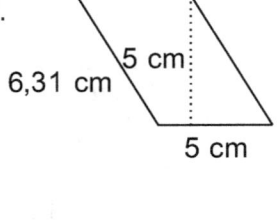

6.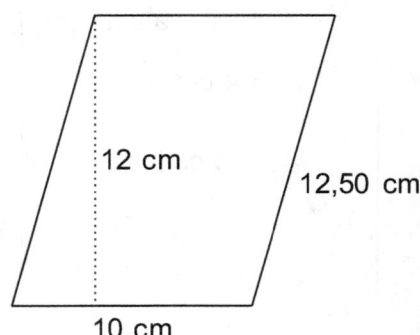

7.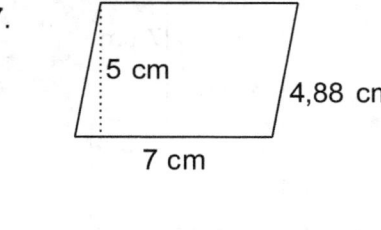

8.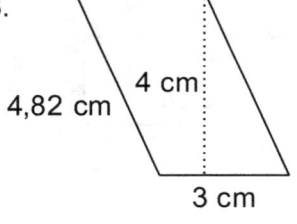

9.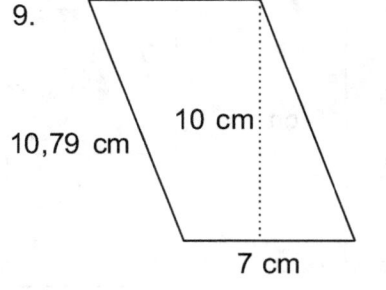

10.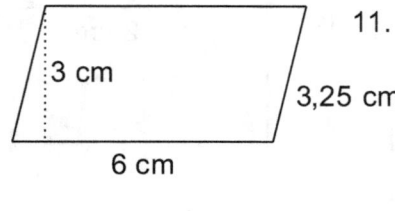

11.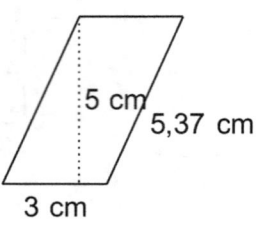

12.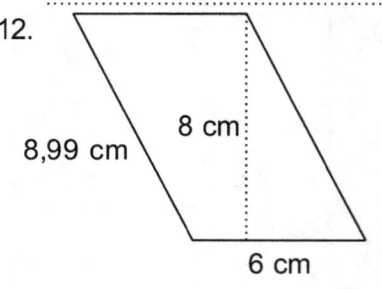

© KingSchool Edition

Find the area.

1. 13 cm, 10 cm

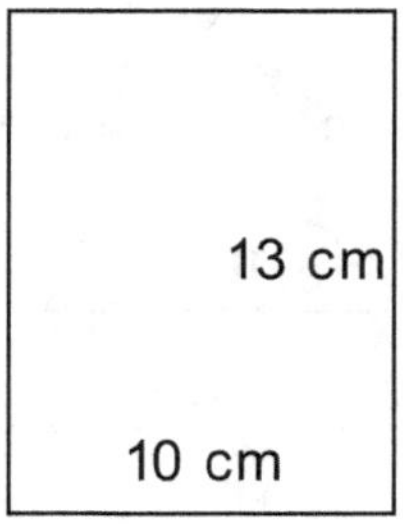

2. 6 cm, 8 cm

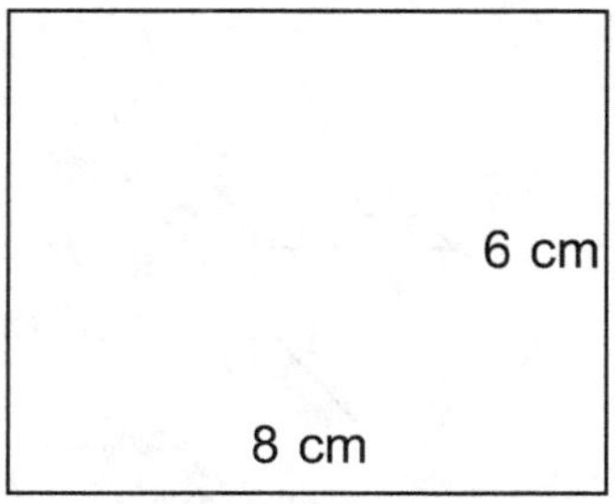

3. 6 cm, 7 cm

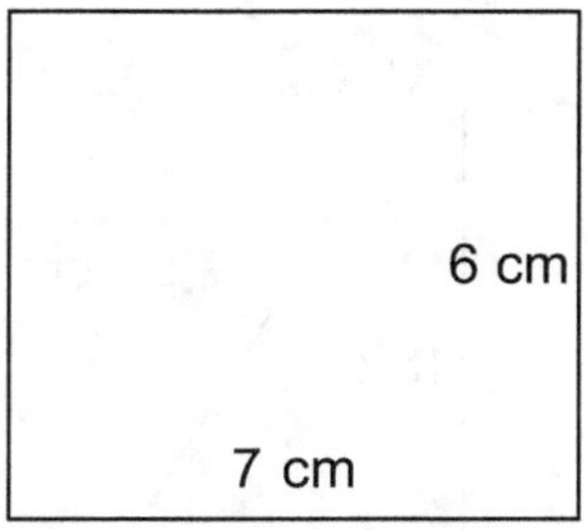

4. 8 cm, 9 cm

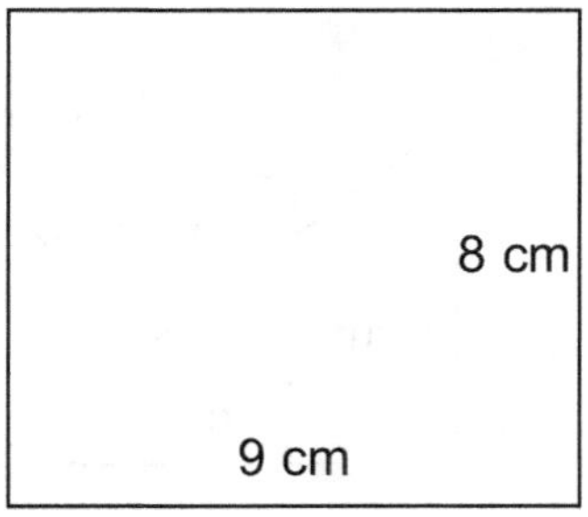

5. 14 cm, 11 cm

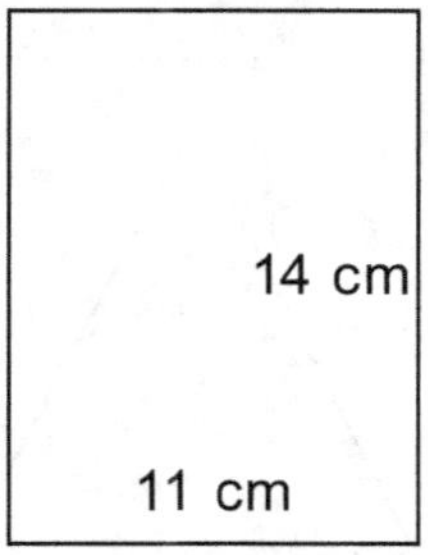

6. 15 cm, 14 cm

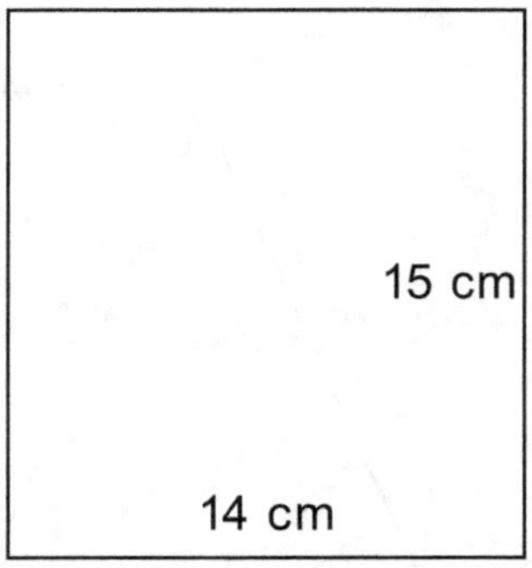

7. 14 cm, 13 cm

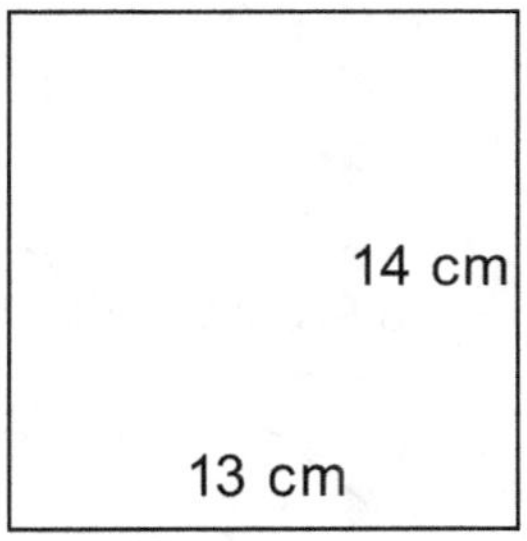

8. 7 cm, 4 cm

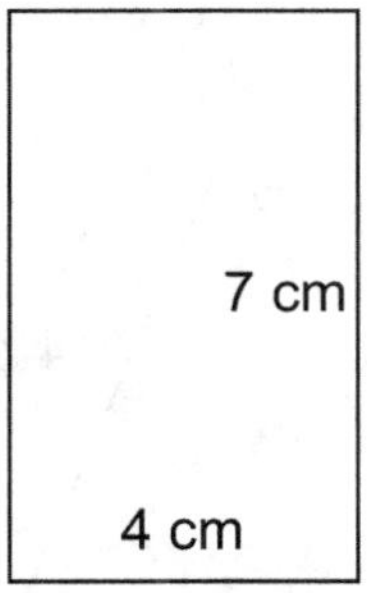

9. 11 cm, 11 cm

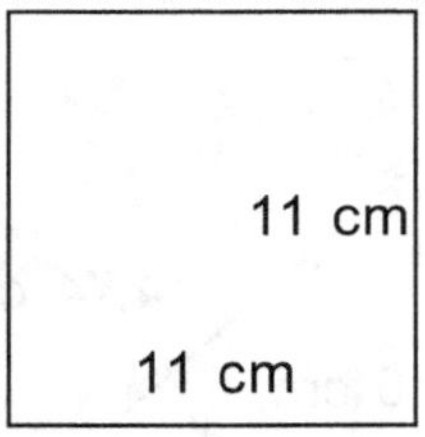

10. 13 cm, 12 cm

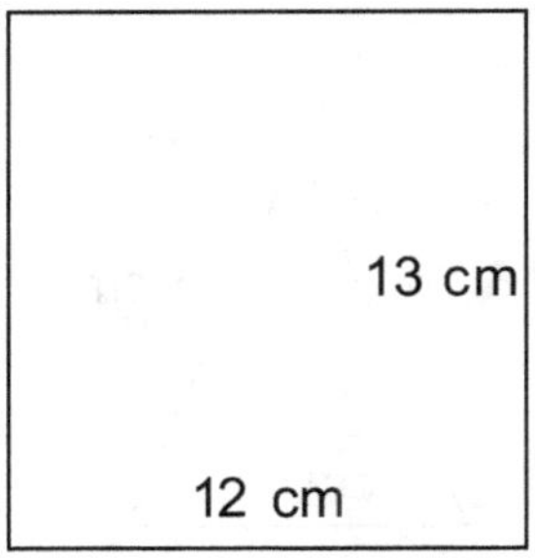

11. 13 cm, 13 cm

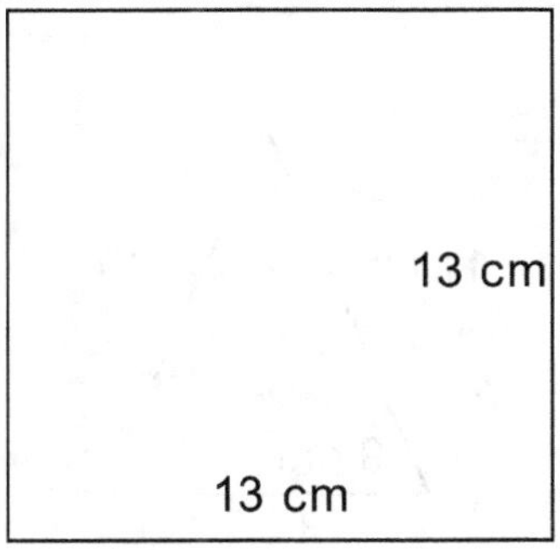

12. 8 cm, 8 cm

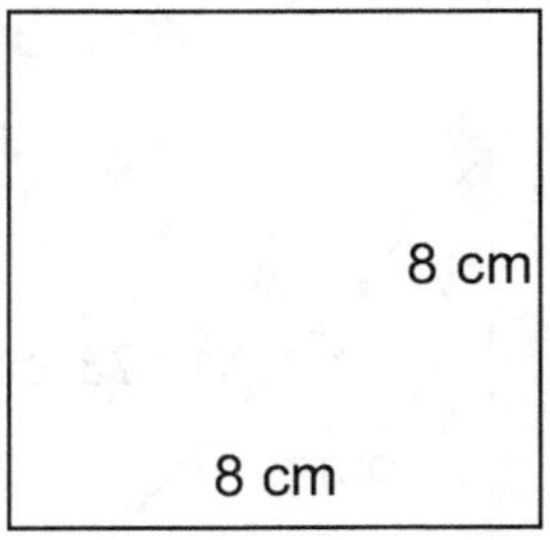

Find the area.

1.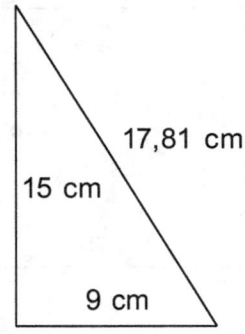

2.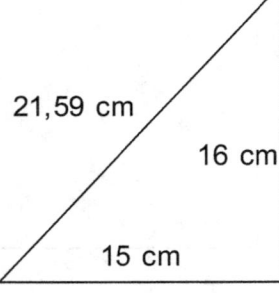

3.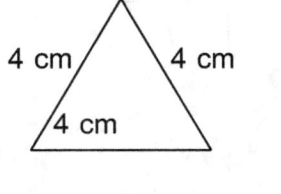

4.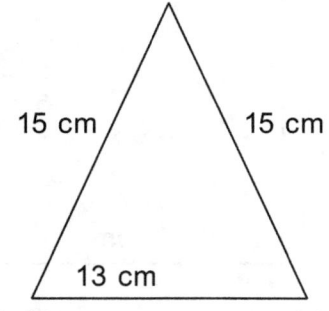

5.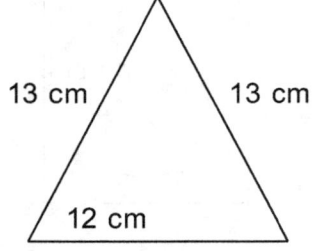

6.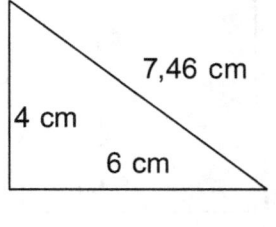

7.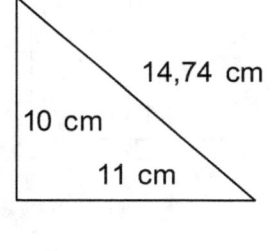

8.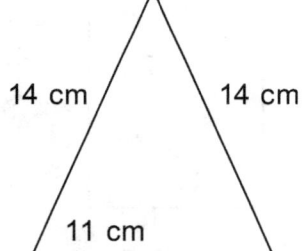

9.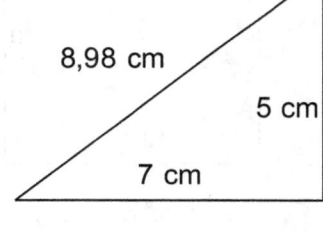

10.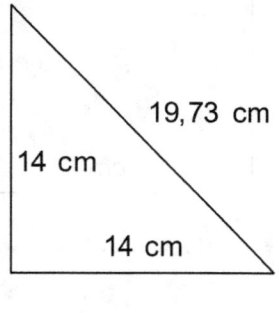

11.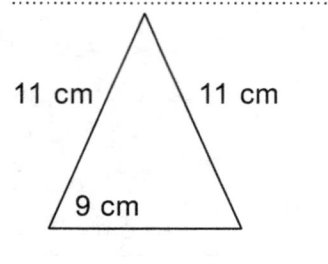

12.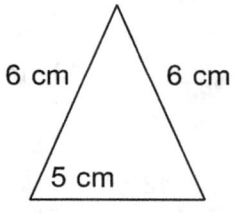

© KingSchool Edition

Find the area.

1.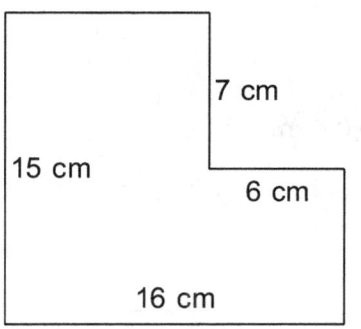

2.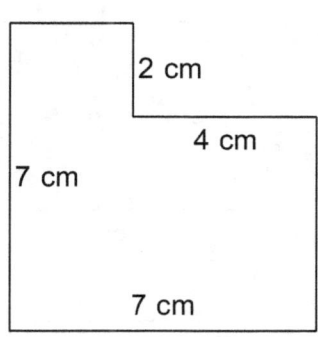

3.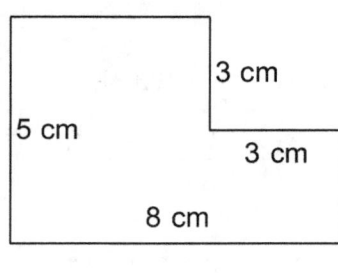

4.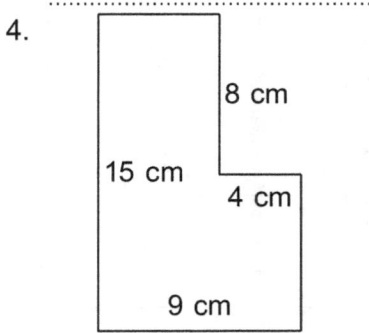

5.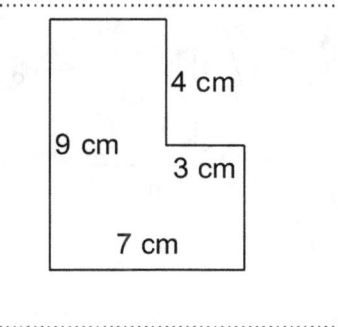

6.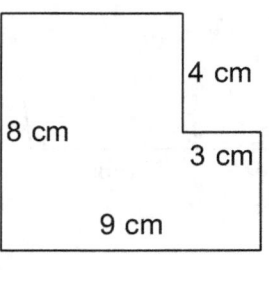

7.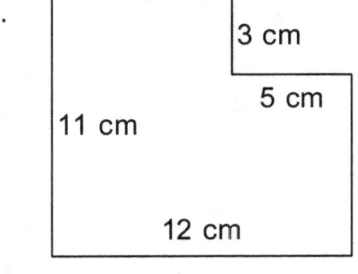

8.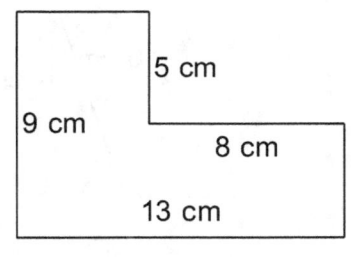

9.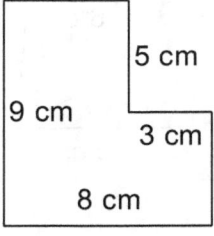

10.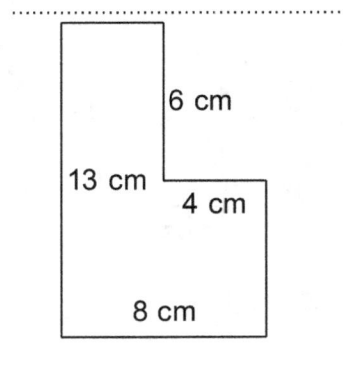

11.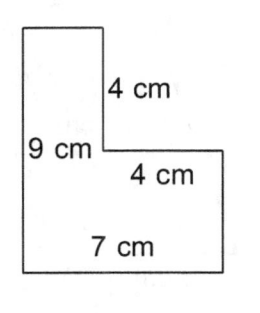

12.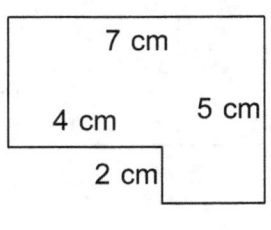

Find the area.

1.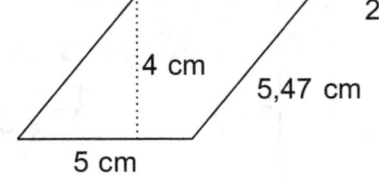
2.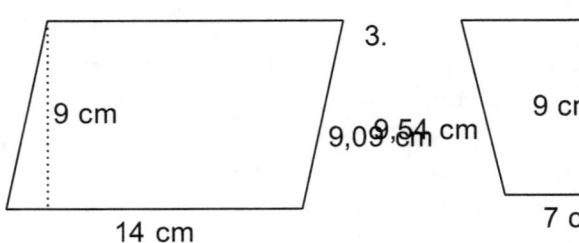
3.
4.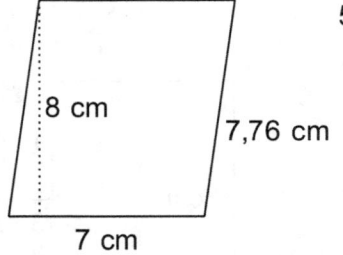
5.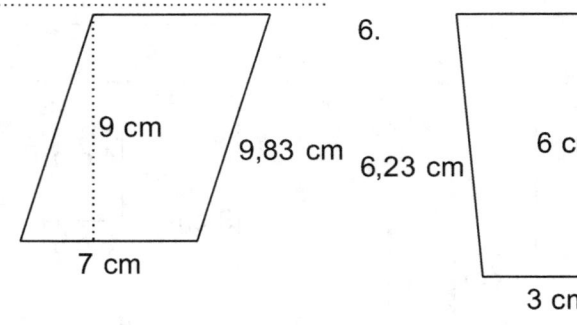
6.
7.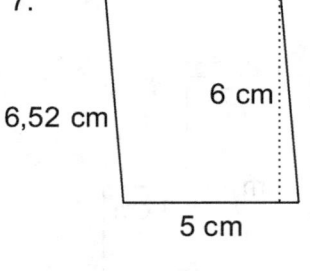
8.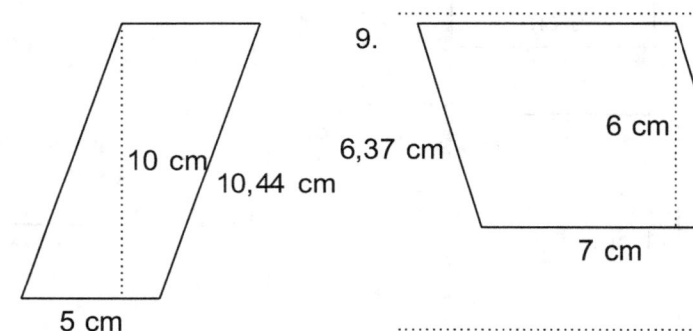
9.
10.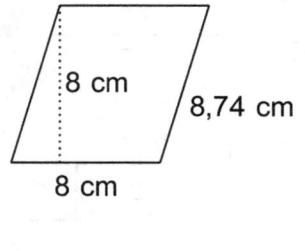
11.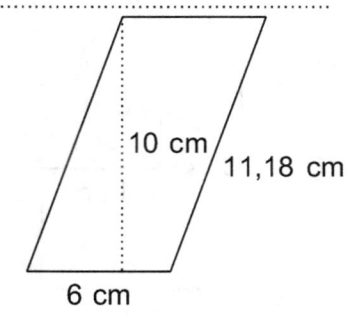
12.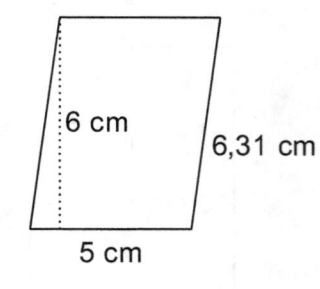

Find the volume.

1.

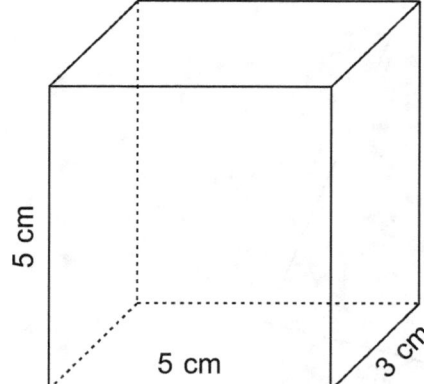

......

2.

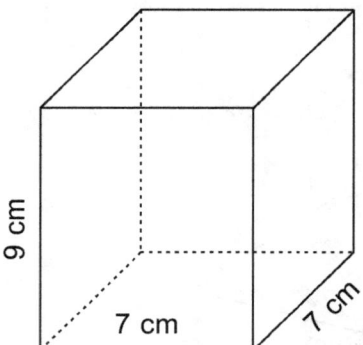

......

3.

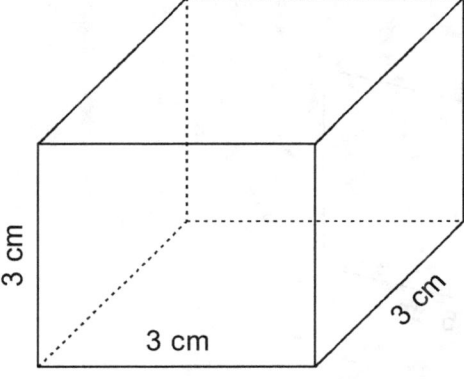

......

Find the volume.

1.

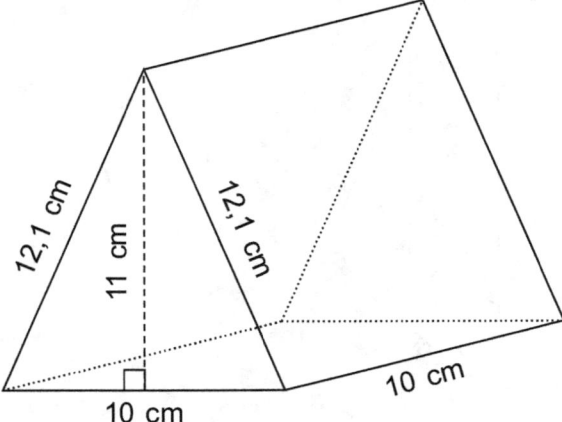

2.

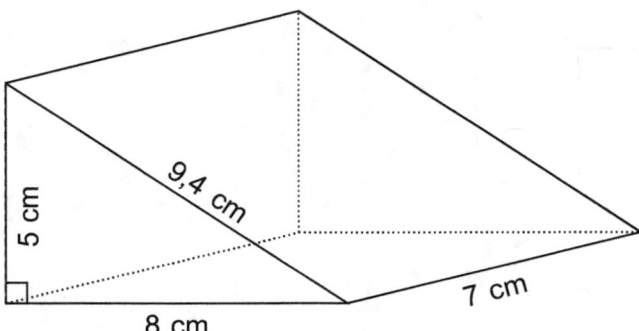

3.

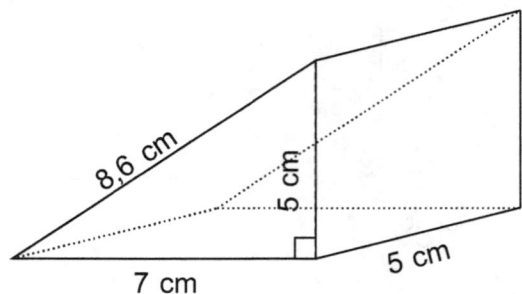

Find the volume.

1.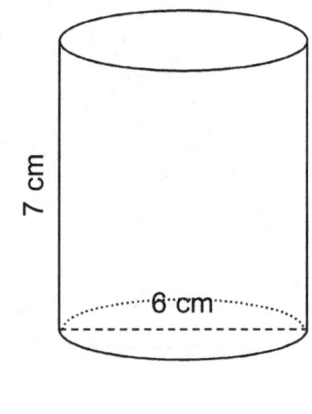

2.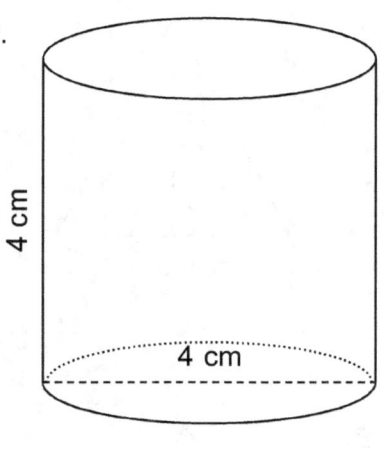

3.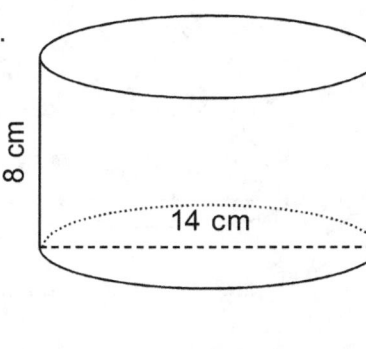

4.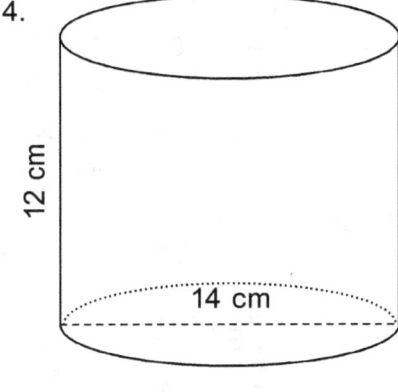

5.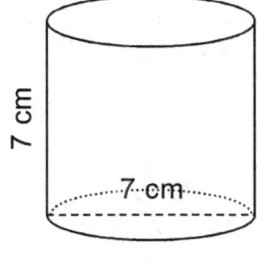

6.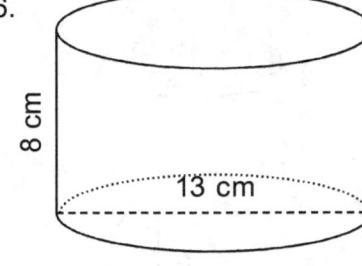

7.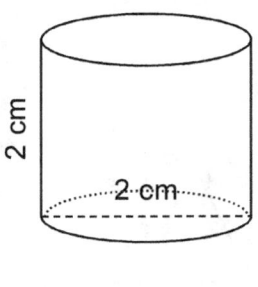

8.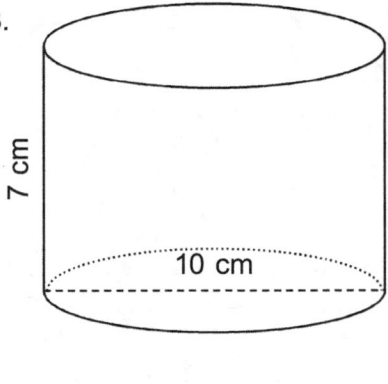

9.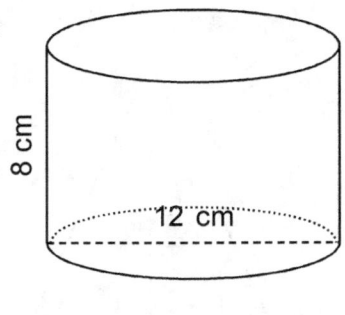

Find the volume.

1.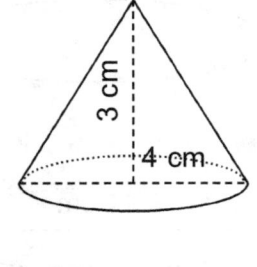

2.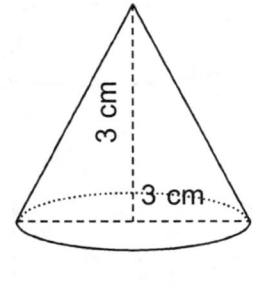

3.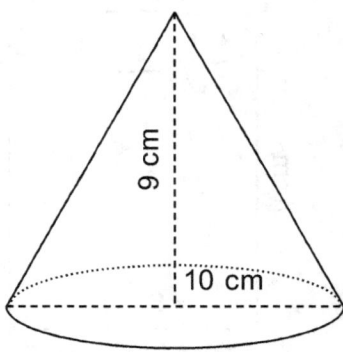

4.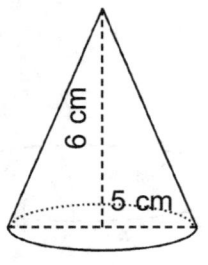

5.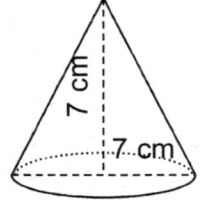

6.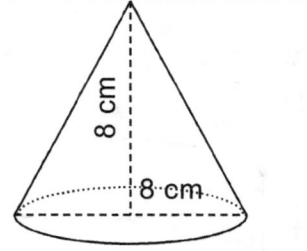

7.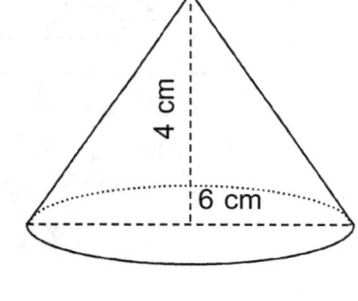

8.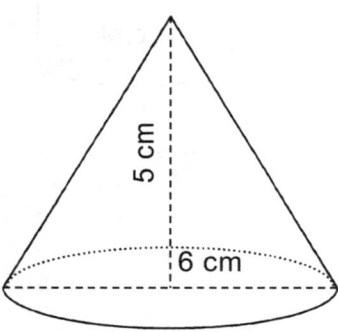

9.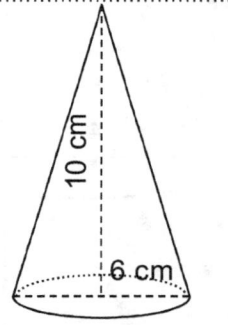

10.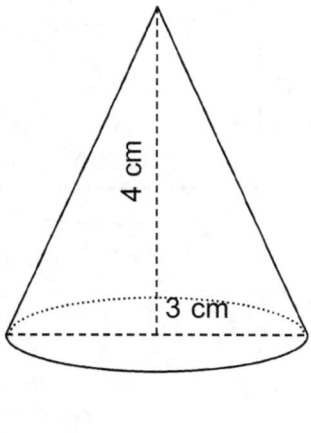

11.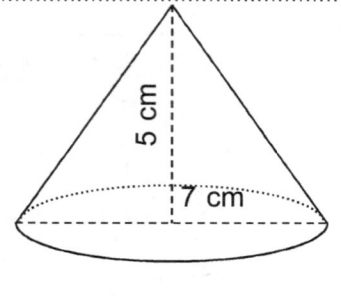

12.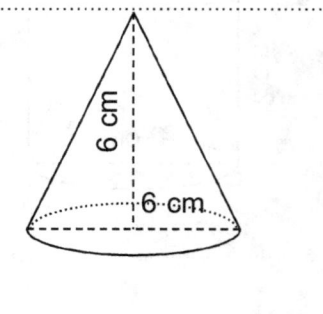

© KingSchool Edition

Find the volume.

1.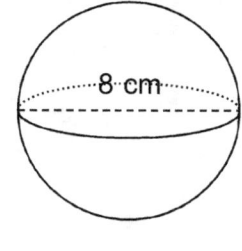
2.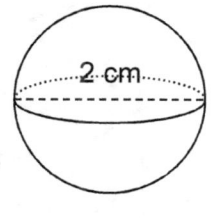
3.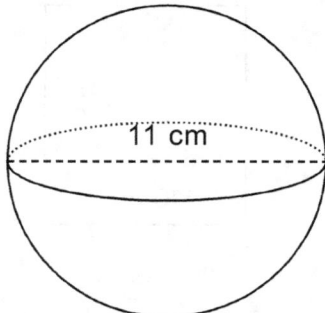

4.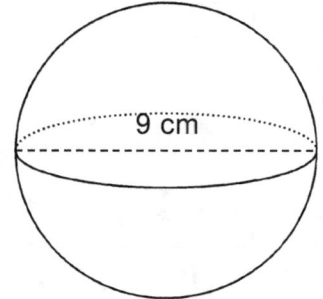
5.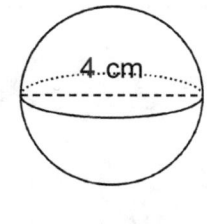
6.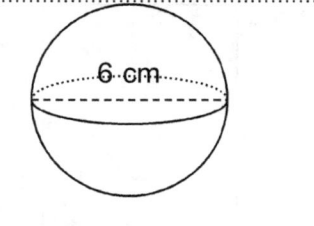

7.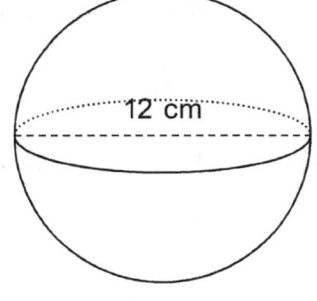
8.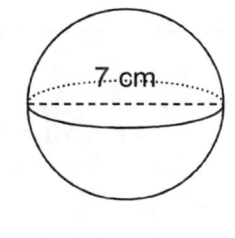
9.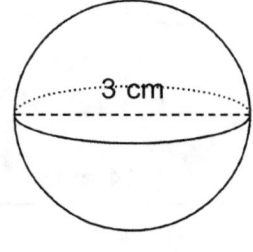

10.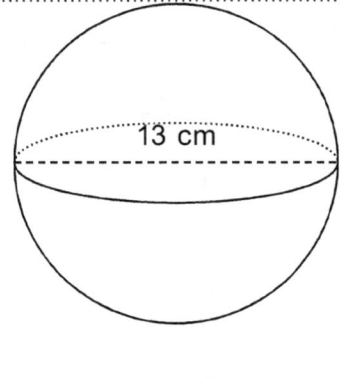
11.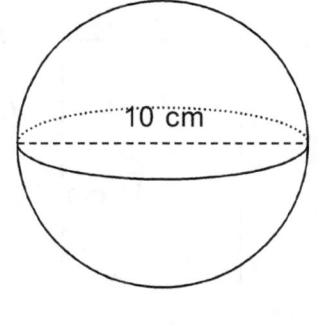
12.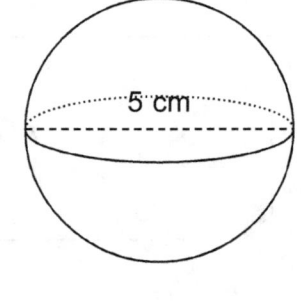

Find the perimeter.

1. 12 cm, 10 cm

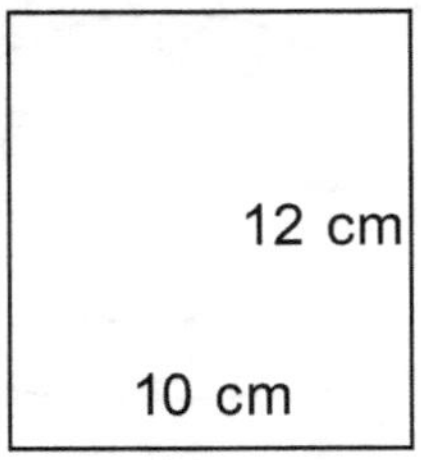

2. 17 cm, 16 cm

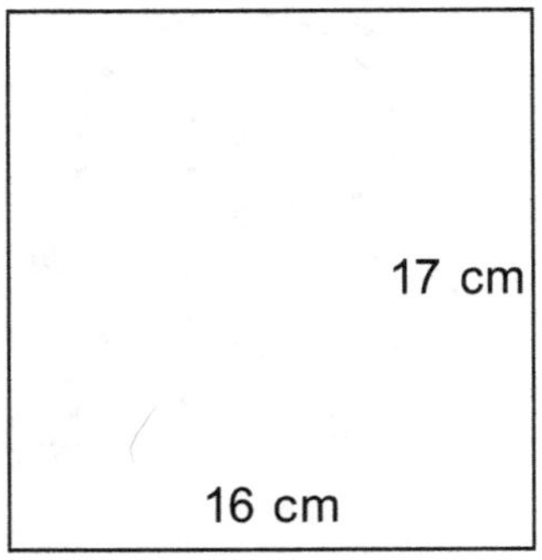

3. 11 cm, 10 cm

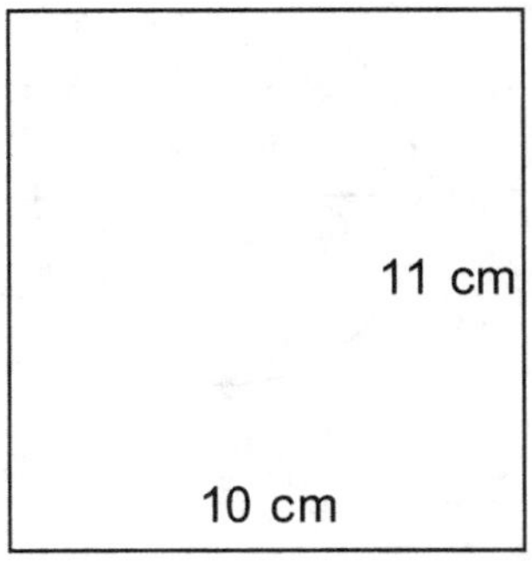

4. 17 cm, 10 cm

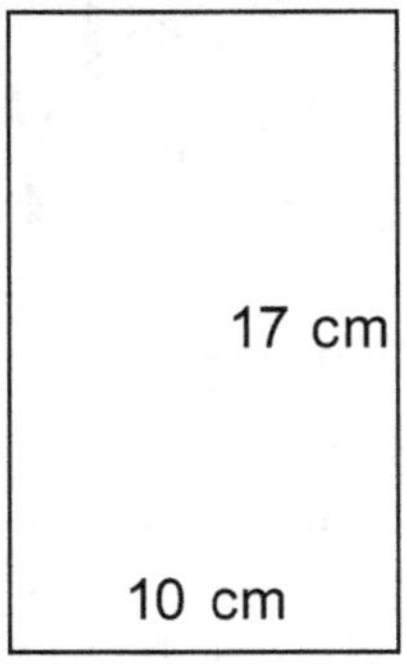

5. 6 cm, 6 cm

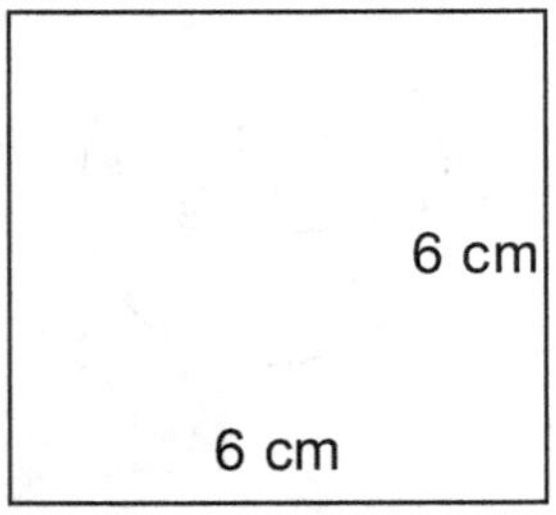

6. 9 cm, 8 cm

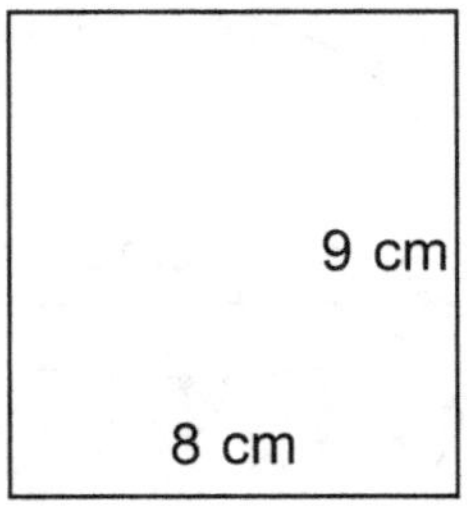

7. 7 cm, 14 cm

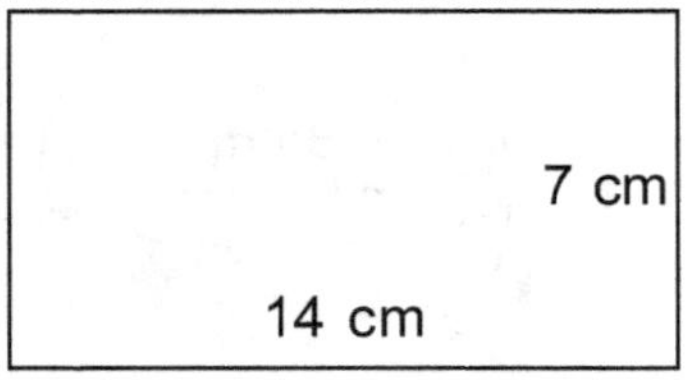

8. 9 cm, 10 cm

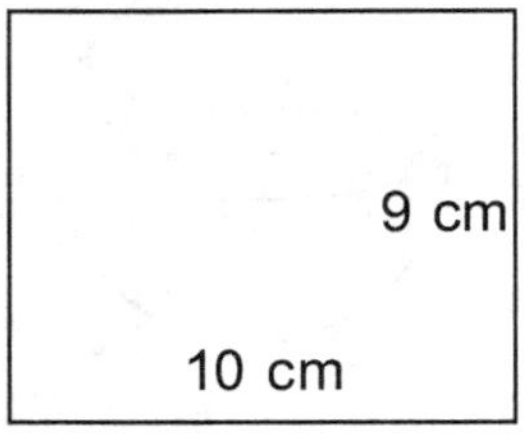

9. 15 cm, 14 cm

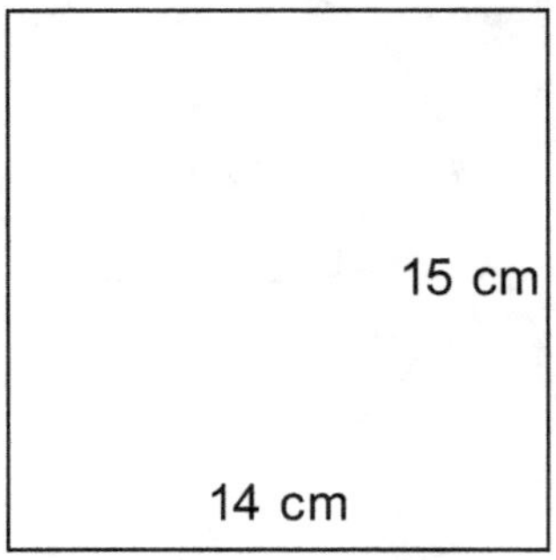

10. 7 cm, 10 cm

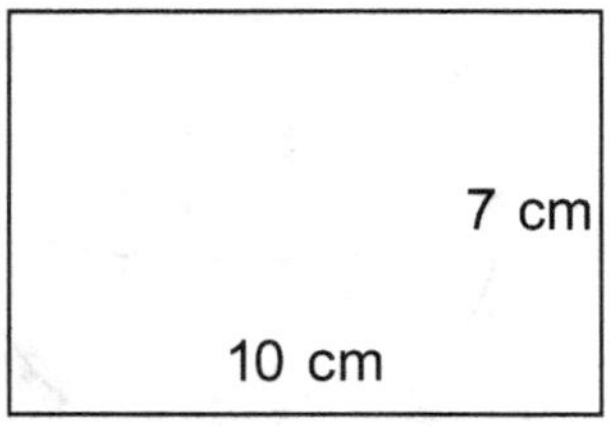

11. 11 cm, 6 cm

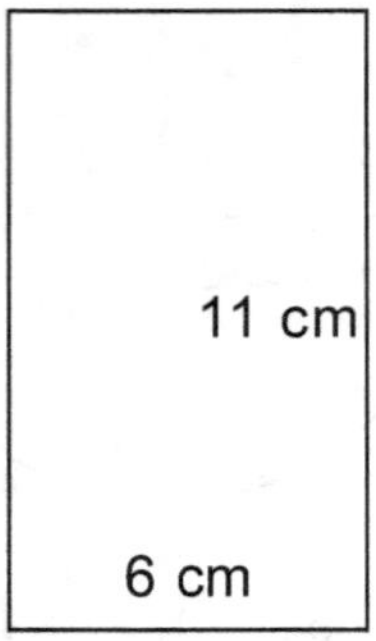

12. 8 cm, 12 cm

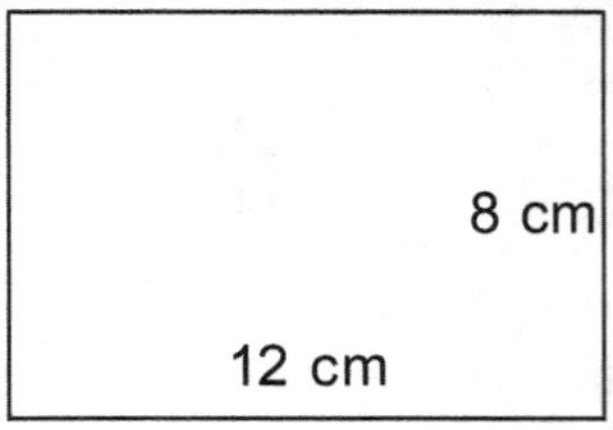

Find the perimeter.

1.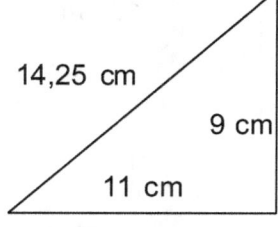
....

2.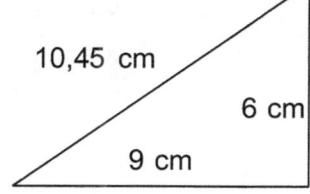
....

3.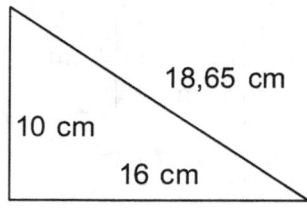
....

4.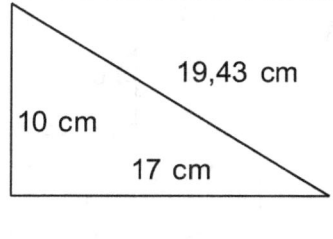
....

5.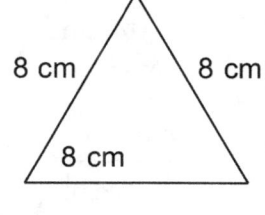
....

6.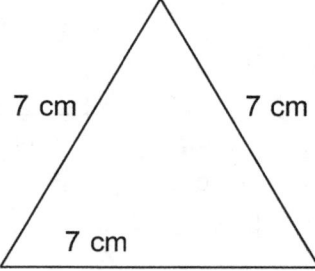
....

7.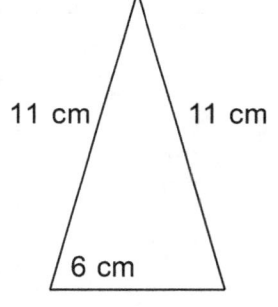
....

8.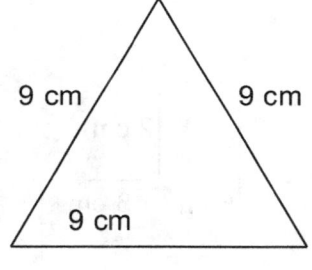
....

9.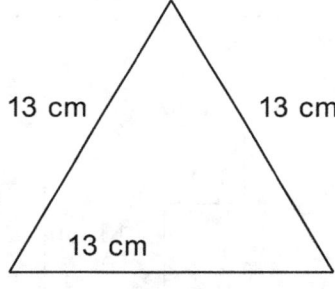
....

10.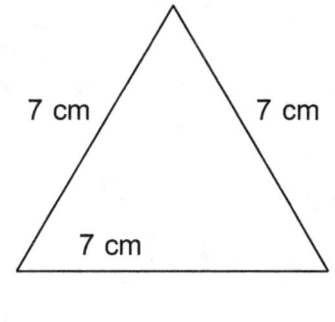
....

11.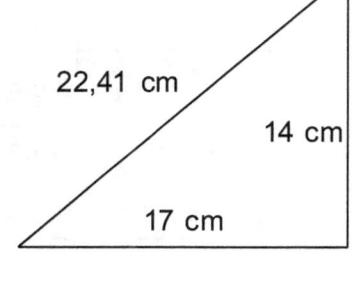
....

12.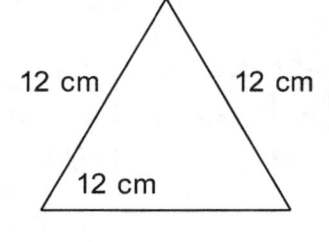
....

Find the perimeter.

1.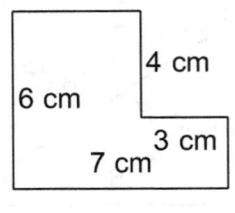
   ....

2.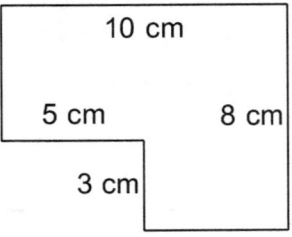
   ....

3.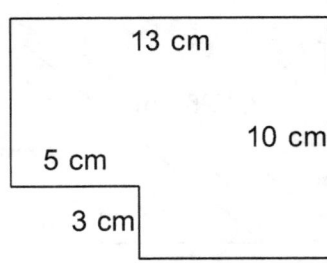
   ....

4.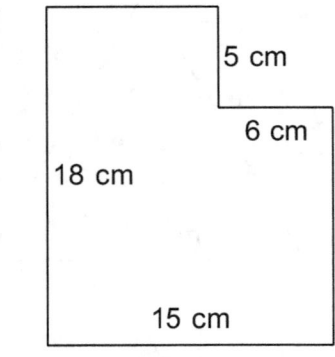
   ....

5.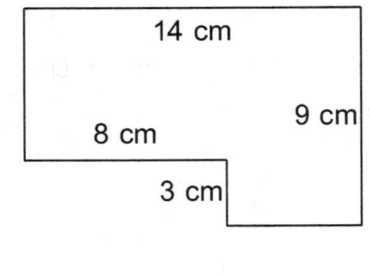
   ....

6.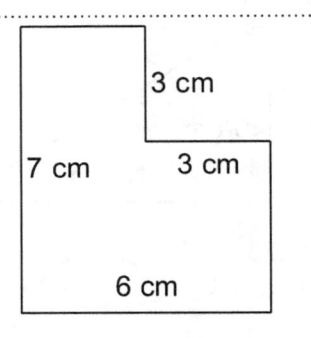
   ....

7.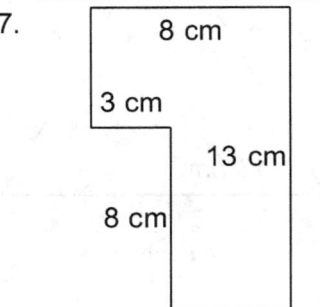
   ....

8.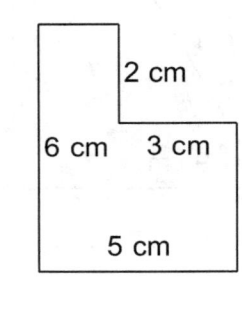
   ....

9.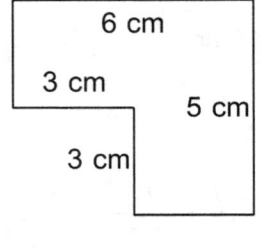
   ....

10.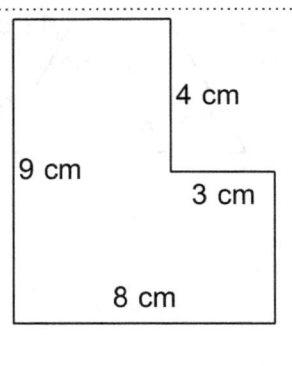
    ....

11.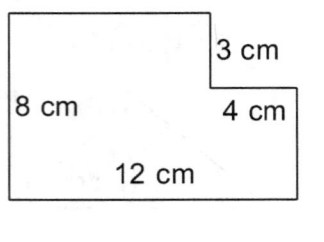
    ....

12.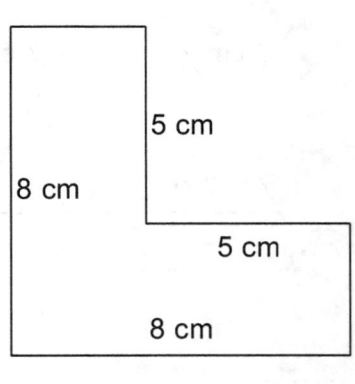
    ....

Find the perimeter.

1.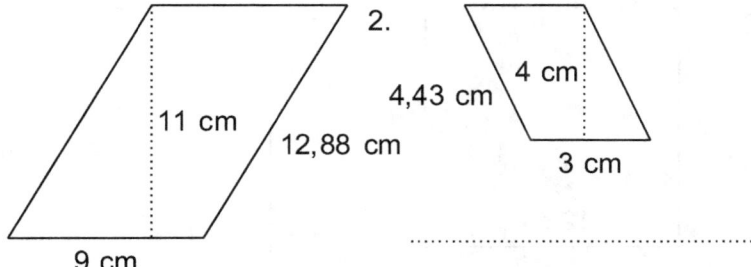

2.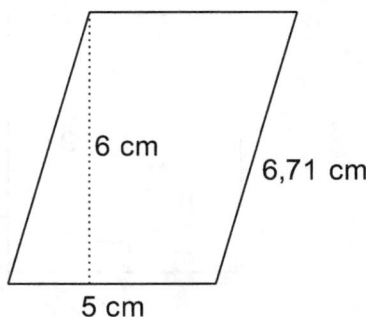

3.

4.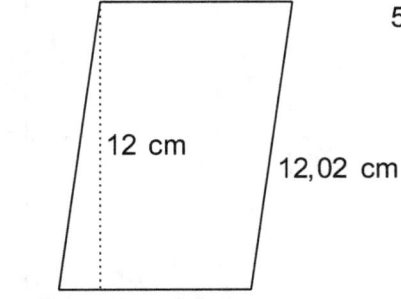

5.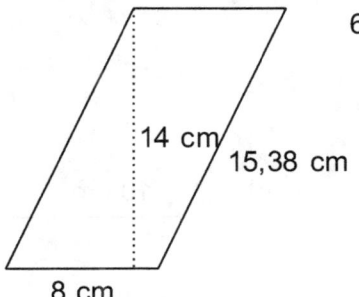

6.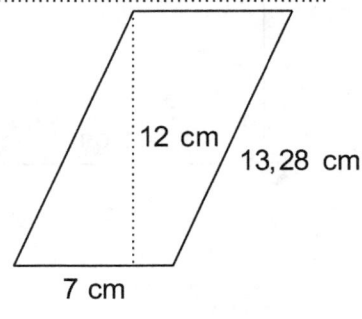

7.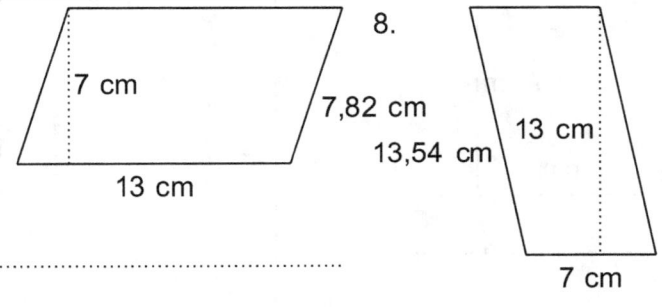

8.

9.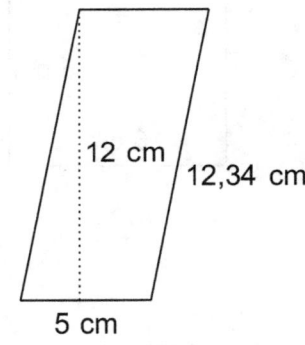

10.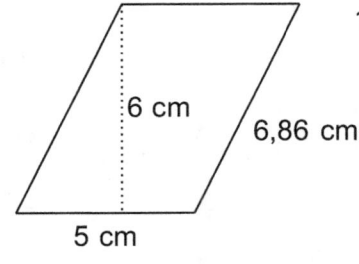

11.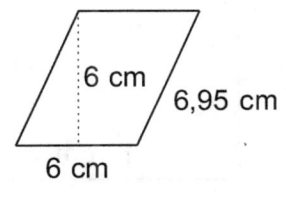

12.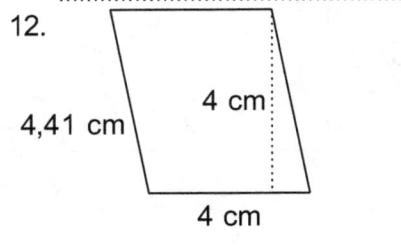

© KingSchool Edition

Find the area.

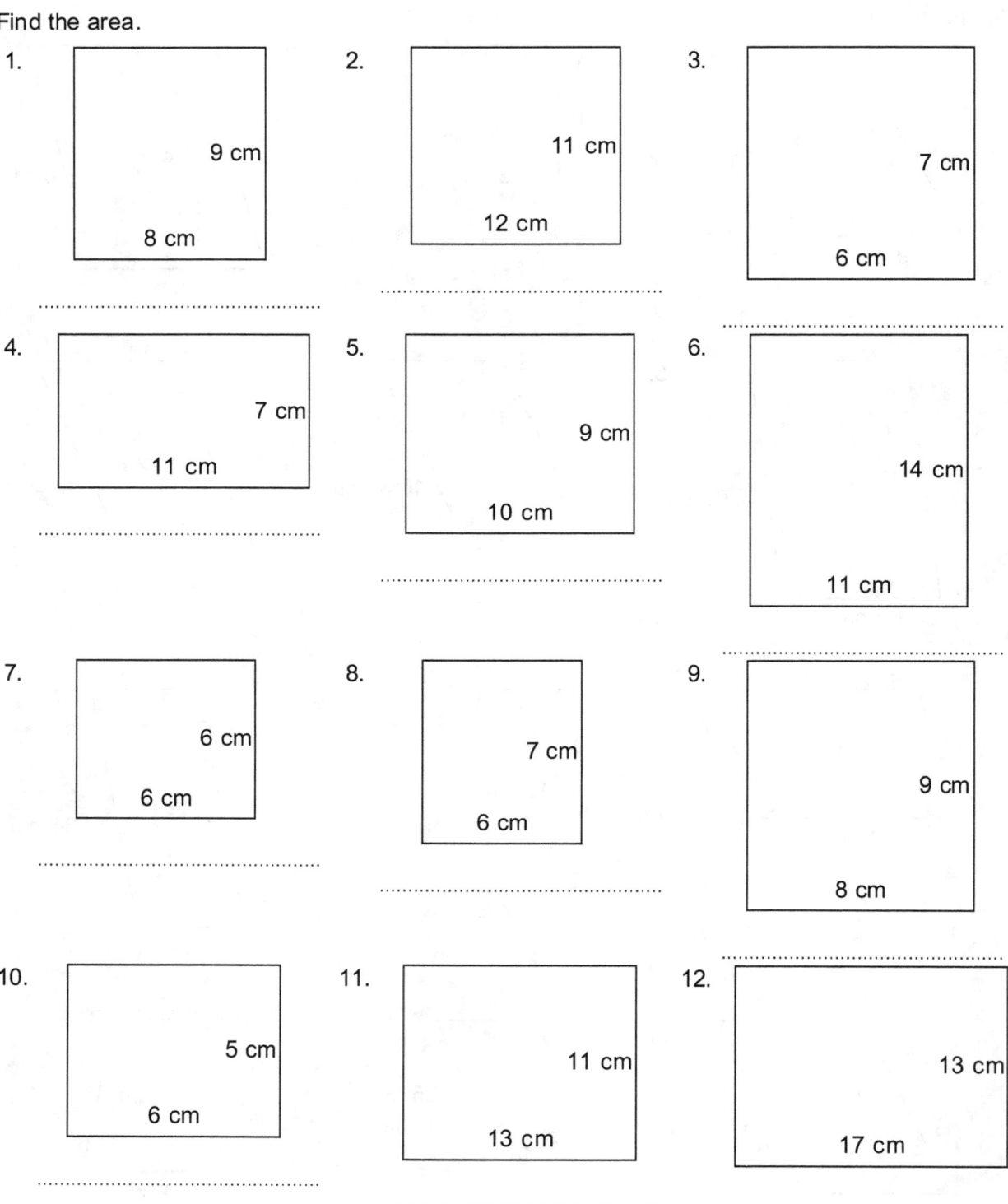

Find the area.

1. 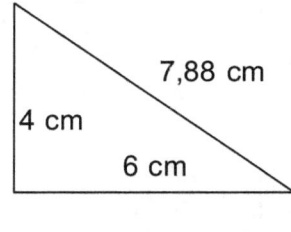 2. 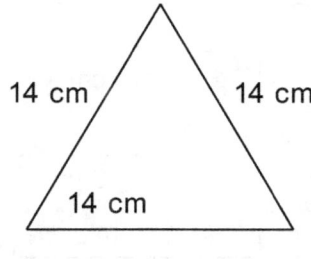 3.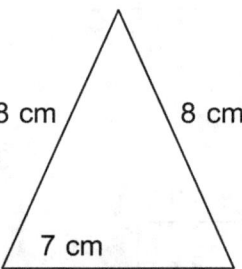

4. 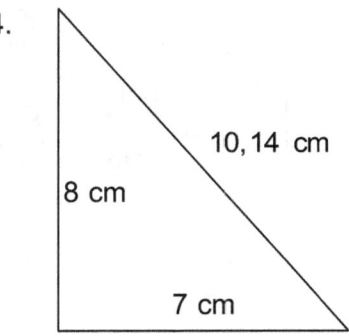 5. 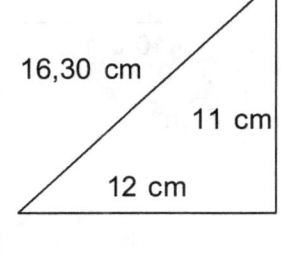 6.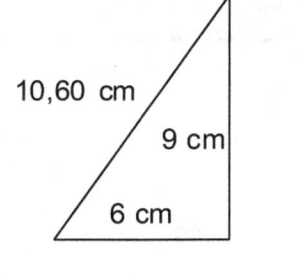

7. 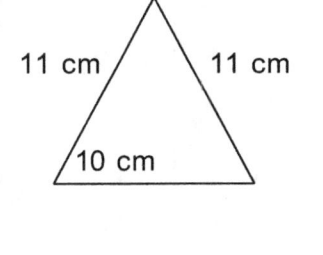 8. 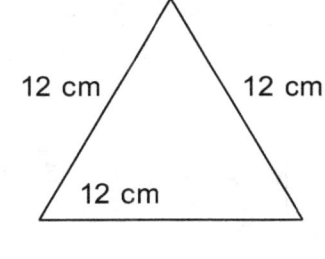 9.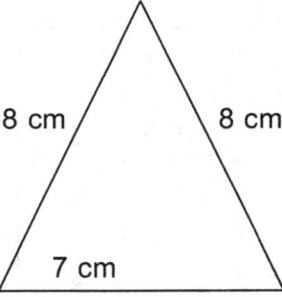

10. 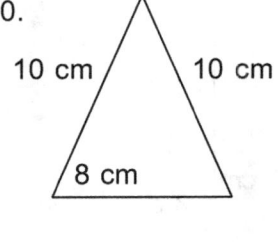 11. 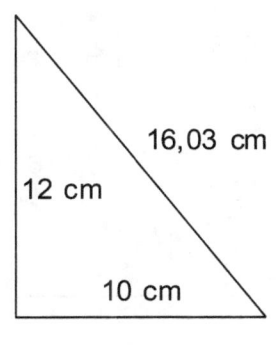 12.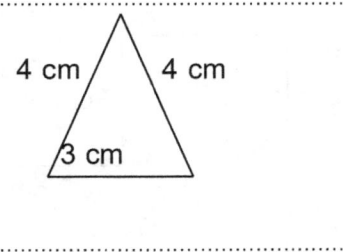

Find the area.

1.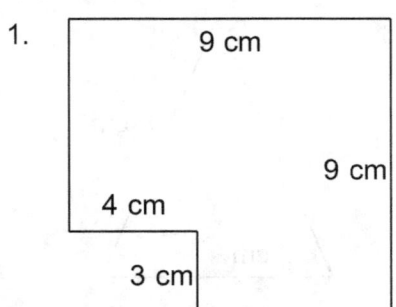
2.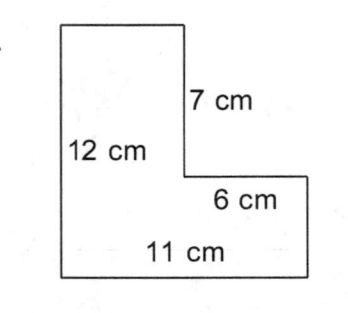
3.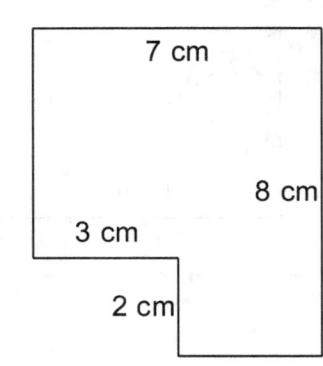
4.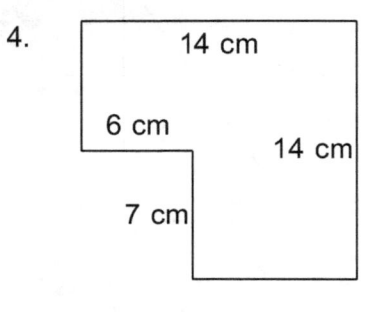
5.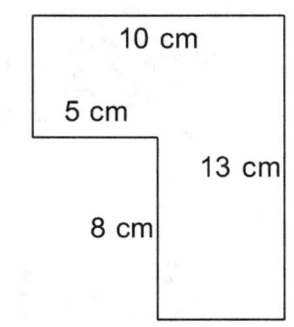
6.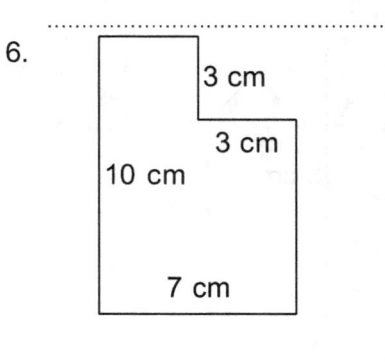
7.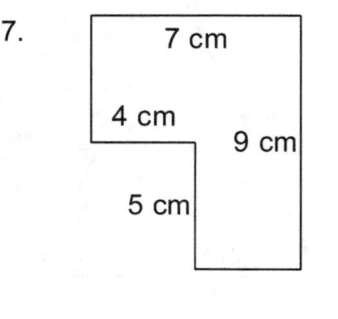
8.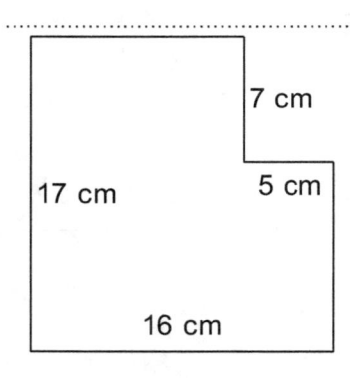
9.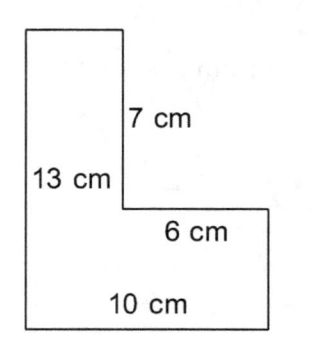
10.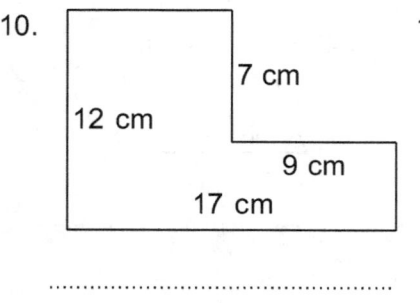
11.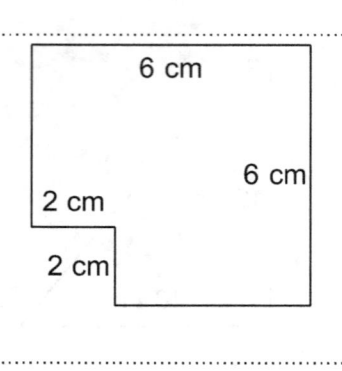
12.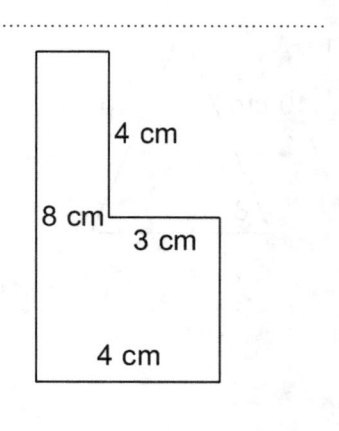

Find the area.

1.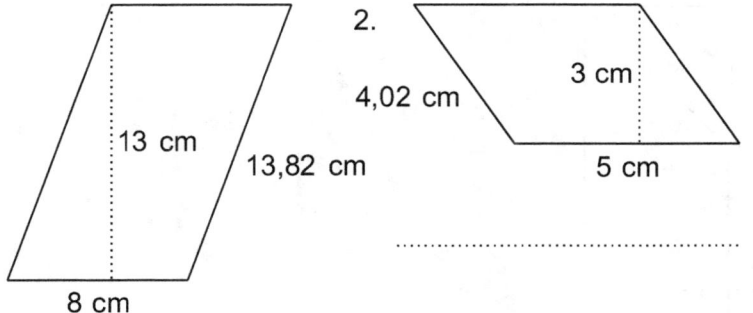
2.
3.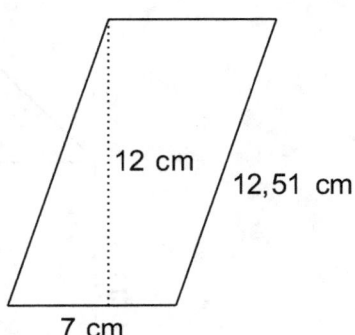

4.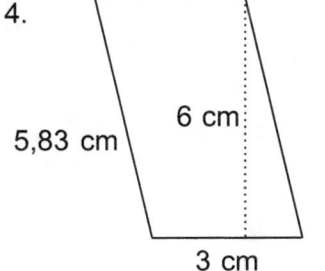
5.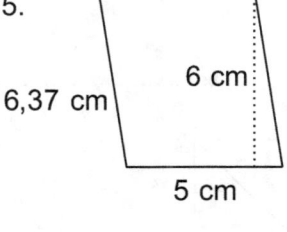
6.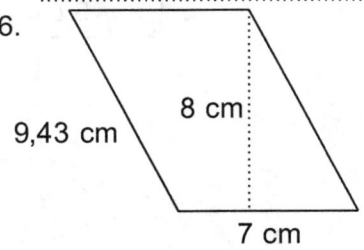

7.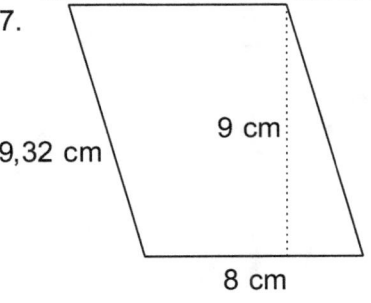
8.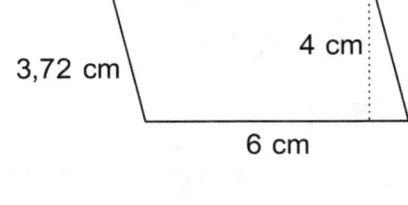
9.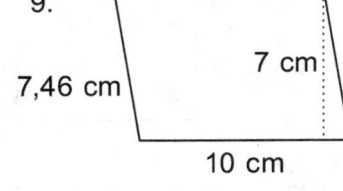

10.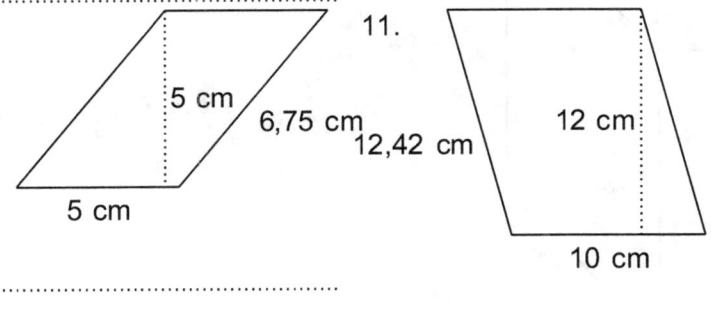
11.
12.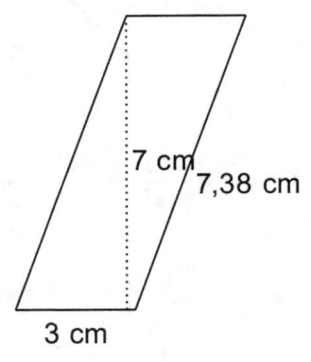

Find the volume.

1.

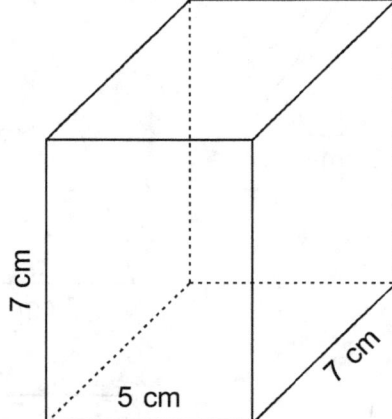

2.

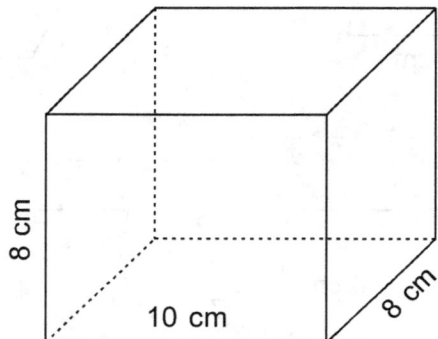

3.

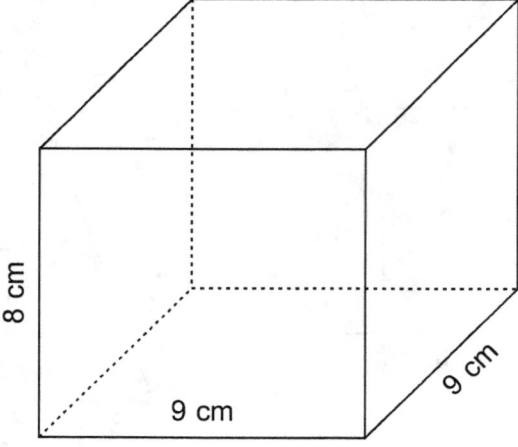

Find the volume.

1.

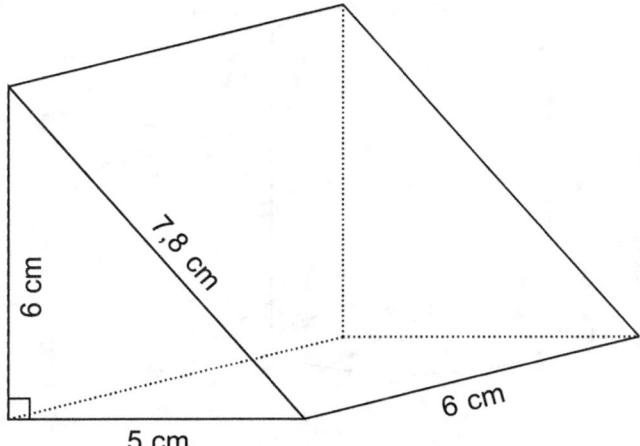

2.

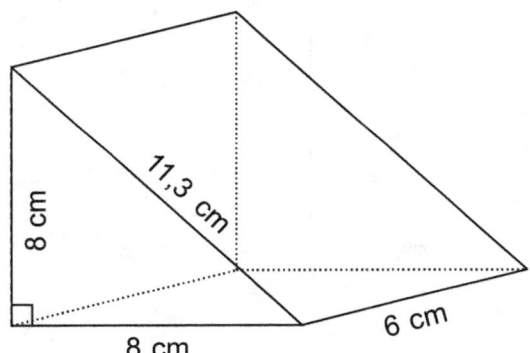

3.

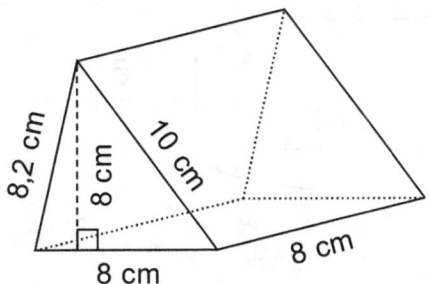

Find the volume.

1.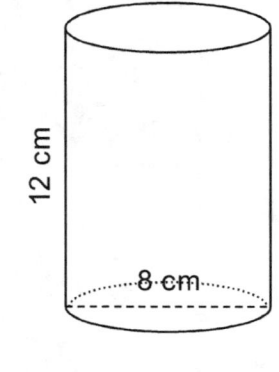

2.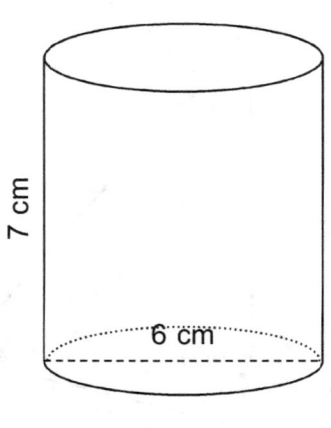

3.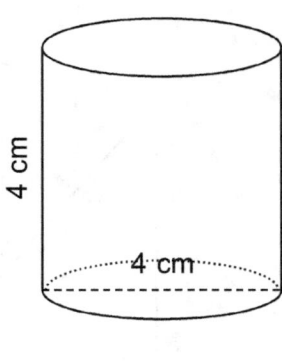

4.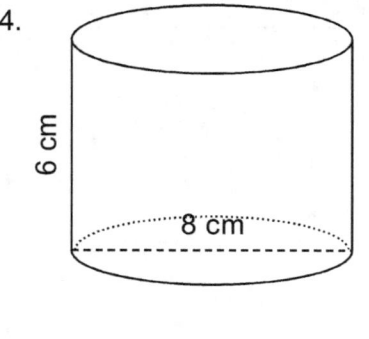

5.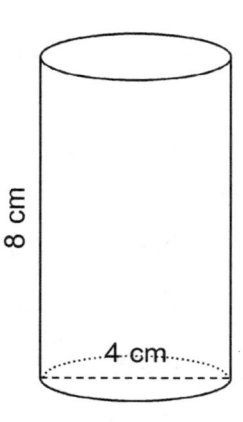

6.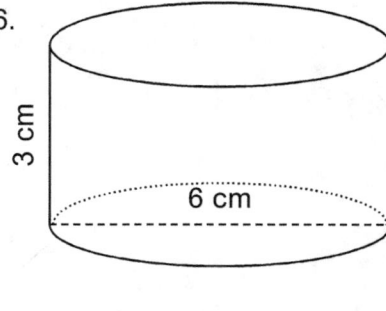

7.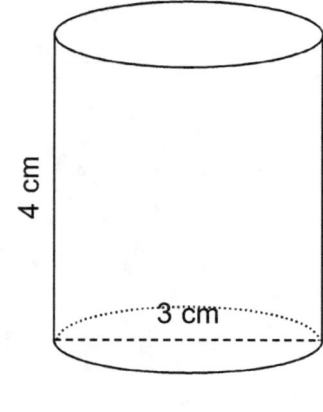

8.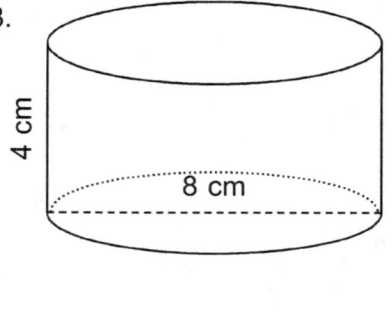

9.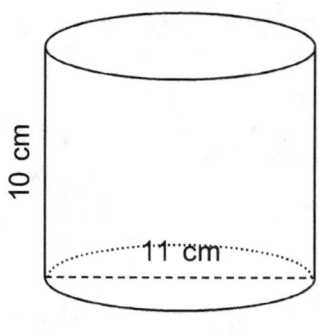

Find the volume.

1.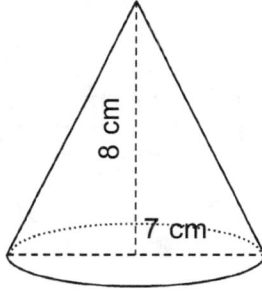

2.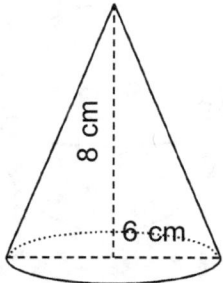

3.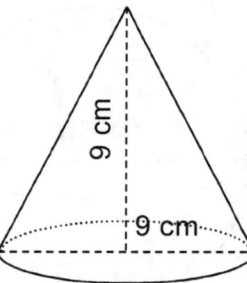

4.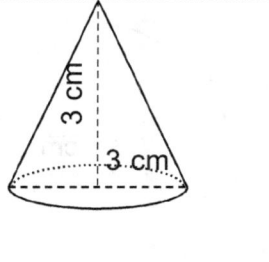

5.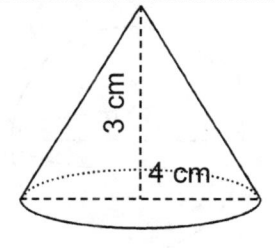

6.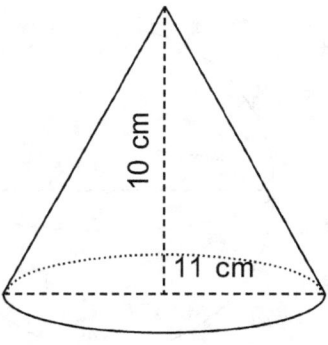

7.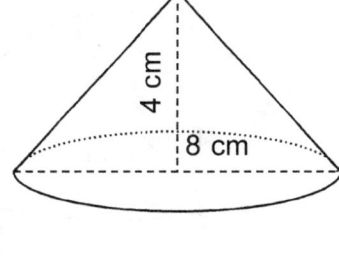

8.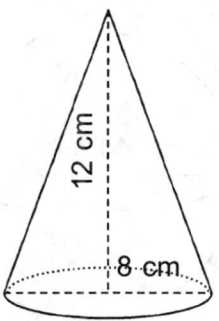

9.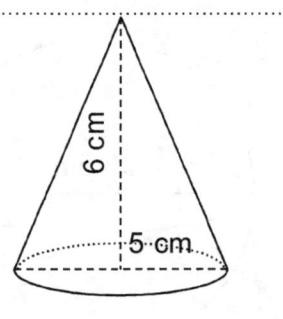

10.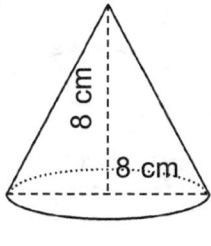

11.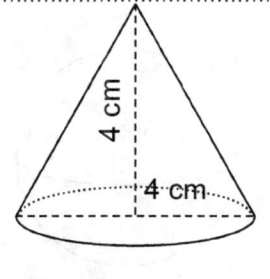

12.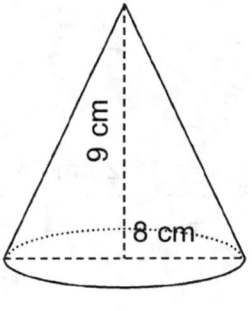

Find the volume.

1.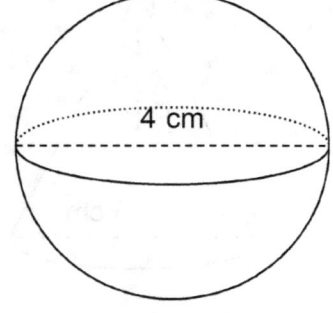

2.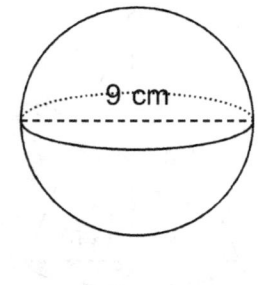

3.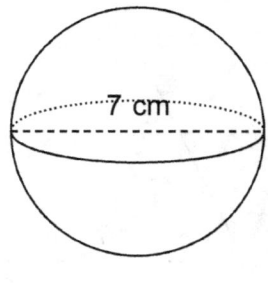

4.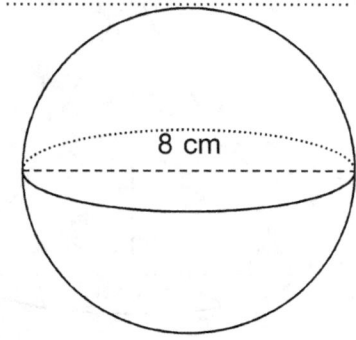

5.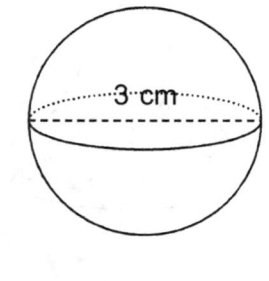

6.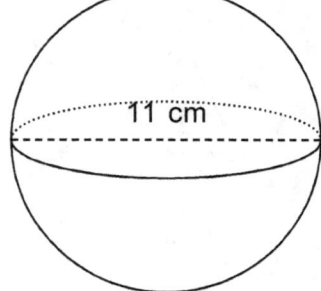

7.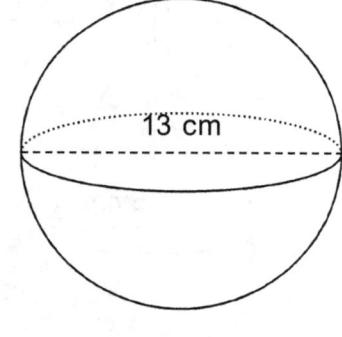

8.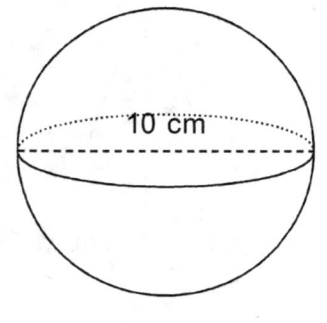

9.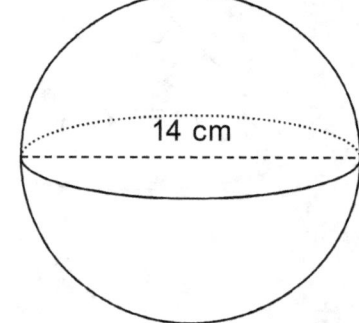

10.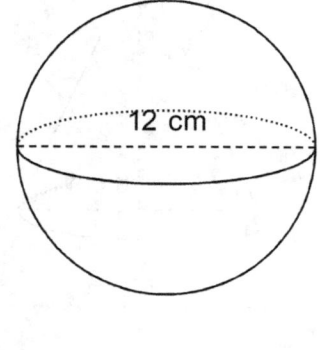

11.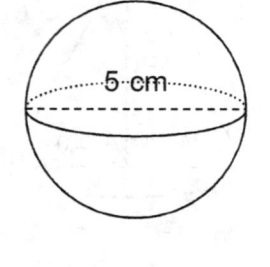

12.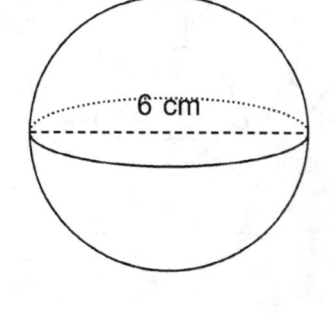

Find the perimeter.

1. 8 cm, 5 cm

2. 6 cm, 7 cm
....

3. 9 cm, 13 cm
....

4. 14 cm, 14 cm
....

5. 4 cm, 6 cm
....

6. 12 cm, 12 cm
....

7. 13 cm, 13 cm

8. 10 cm, 13 cm
....

9. 10 cm, 7 cm
....

10. 7 cm, 9 cm
....

11. 6 cm, 4 cm
....

12. 10 cm, 7 cm
....

Find the perimeter.

1.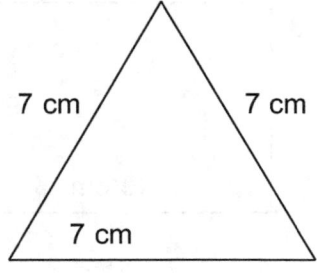

2.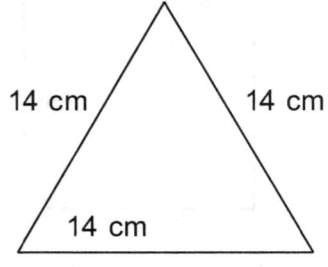

3.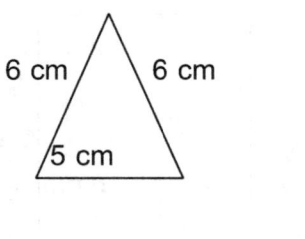

4.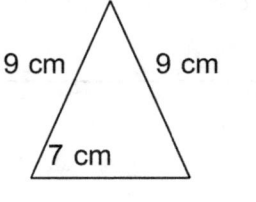

5.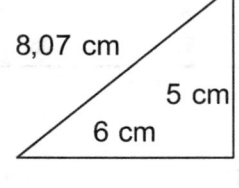

6.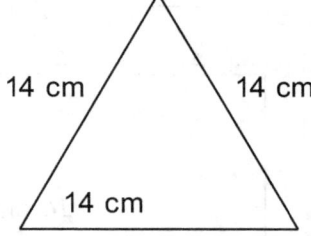

7.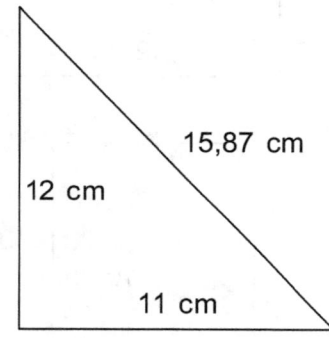

8.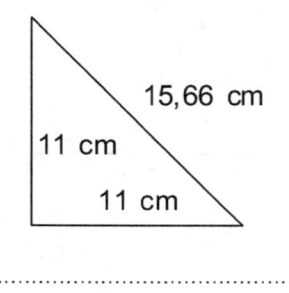

9.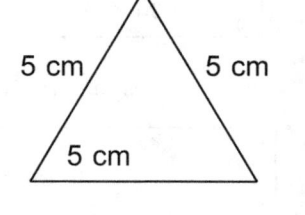

10.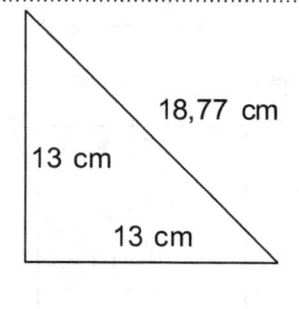

11.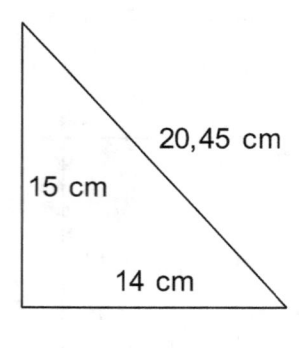

12.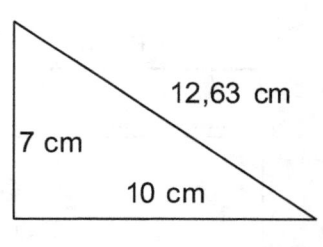

Find the perimeter.

1.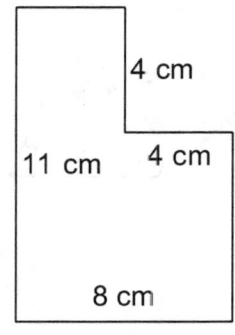

2.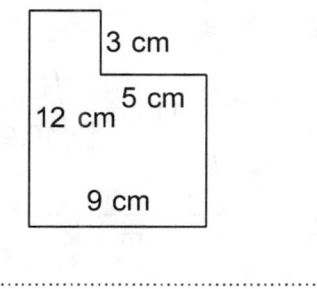

3.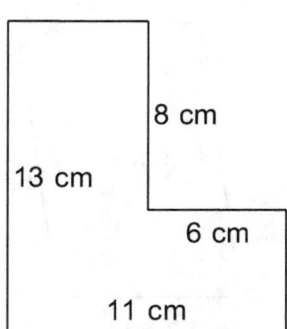

4.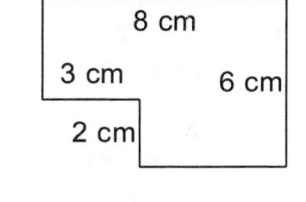

5.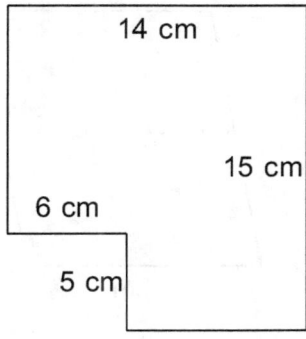

6.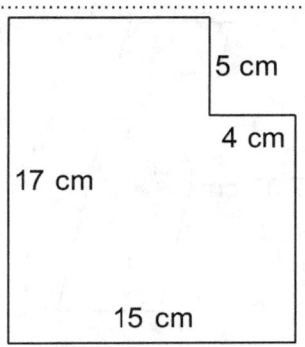

7.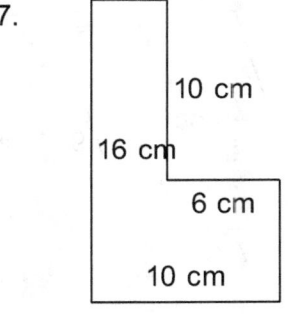

8.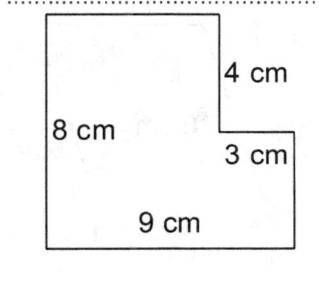

9.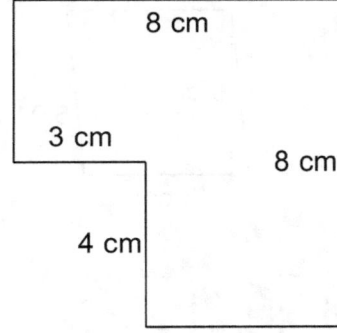

10.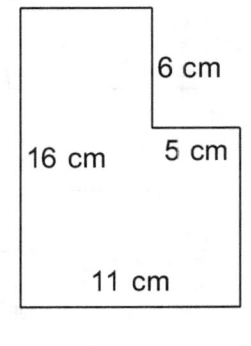

11.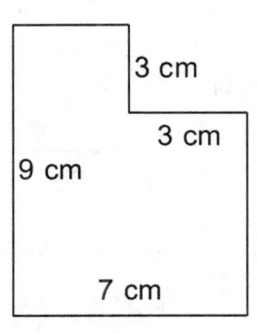

12.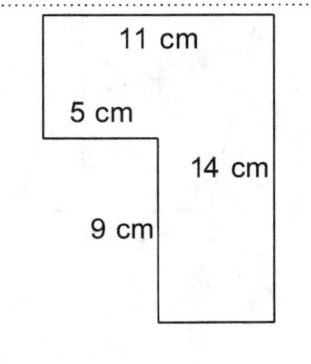

Find the perimeter.

1.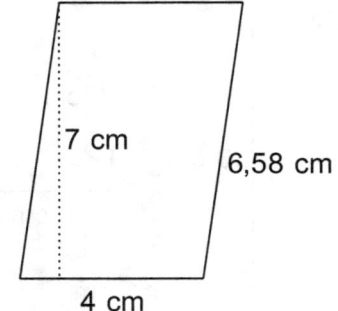

2.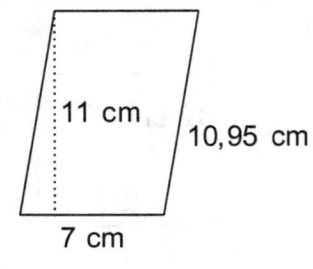

3.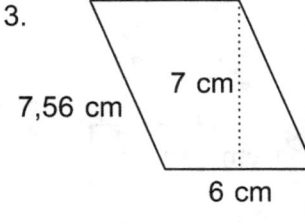

....

4.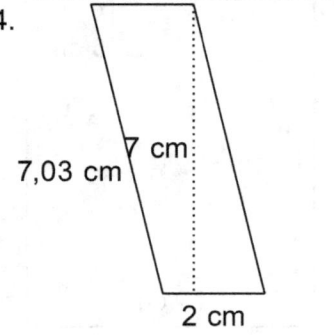

5.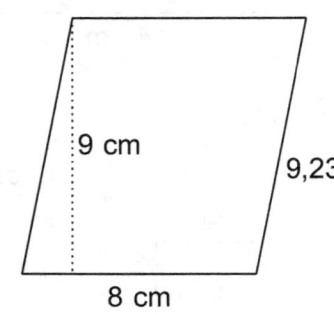

6.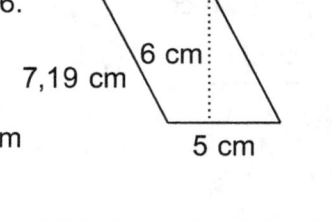

....

7.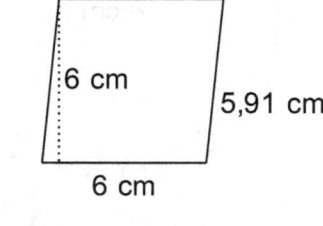

....

8.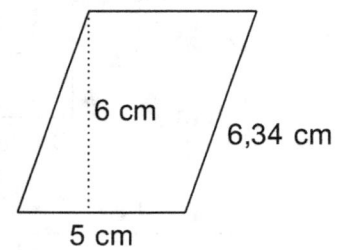

9.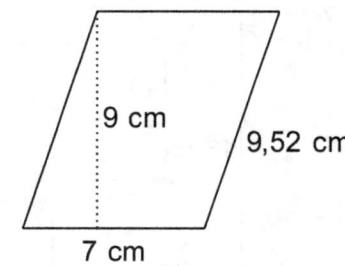

....

10.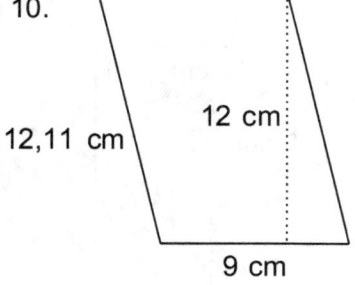

11.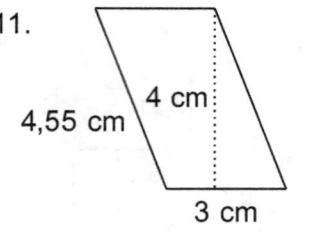

12.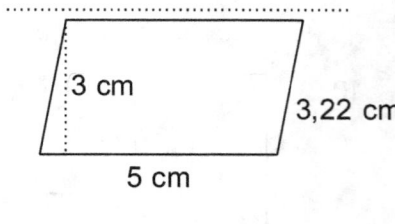

....

© KingSchool Edition

Find the area.

1. 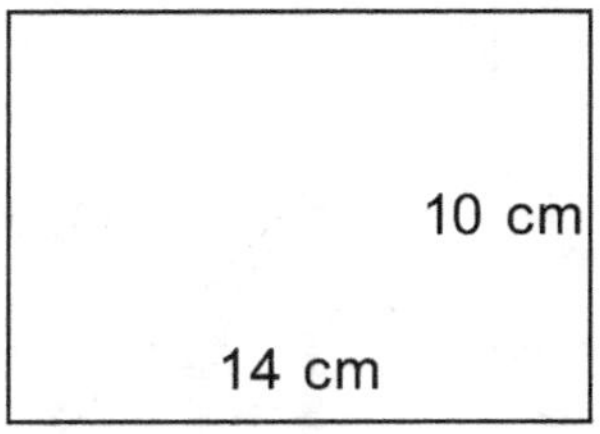 10 cm, 14 cm

2. 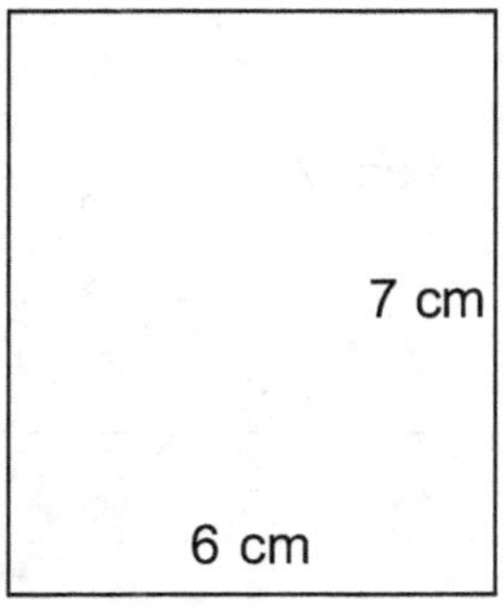 7 cm, 6 cm

3. 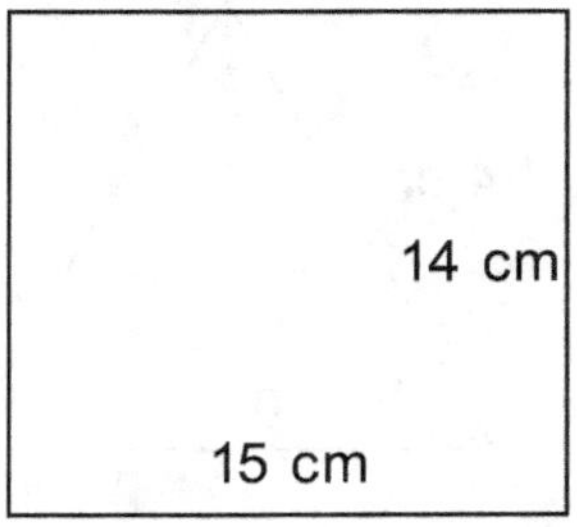 14 cm, 15 cm

4. 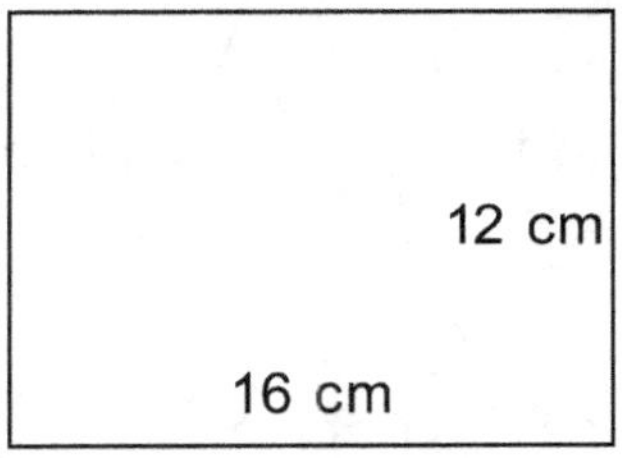 12 cm, 16 cm

5. 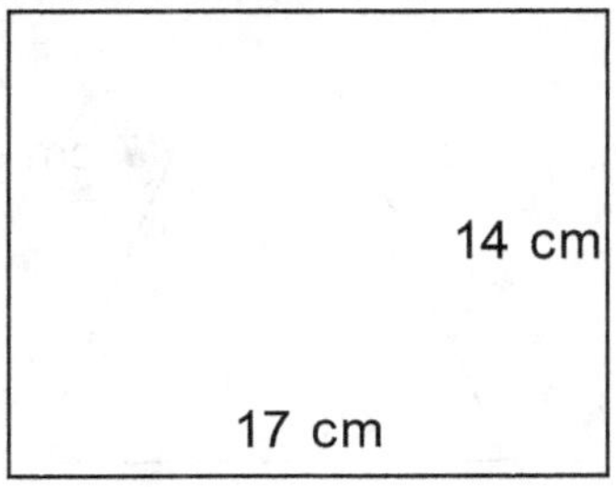 14 cm, 17 cm

6. 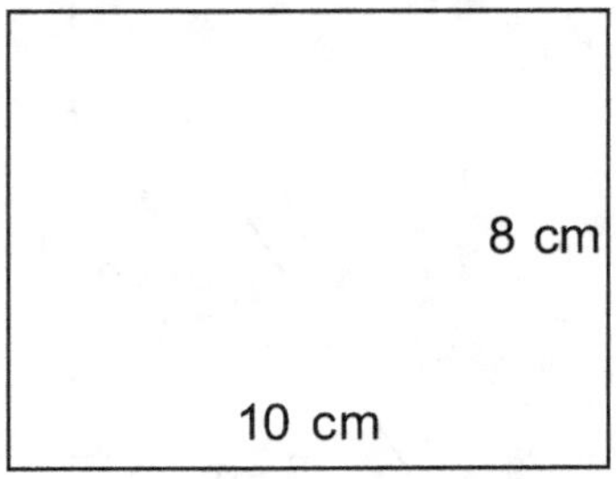 8 cm, 10 cm

7. 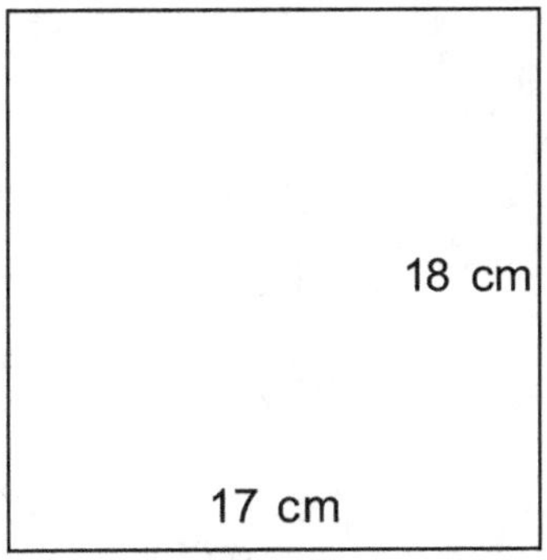 18 cm, 17 cm

8. 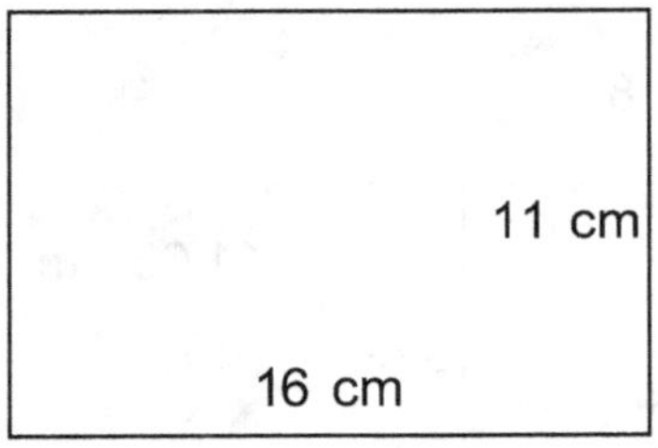 11 cm, 16 cm

9. 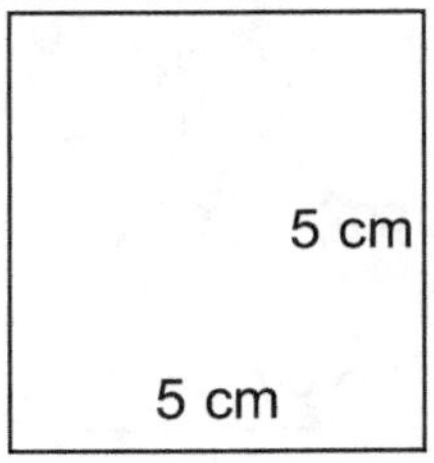 5 cm, 5 cm

10. 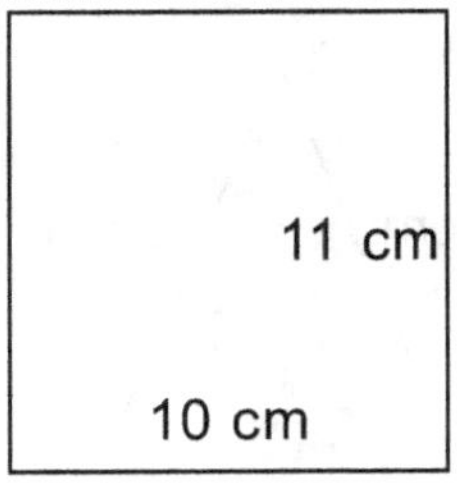 11 cm, 10 cm

11. 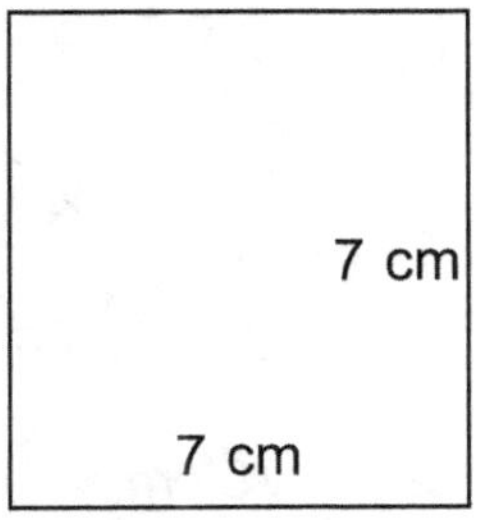 7 cm, 7 cm

12. 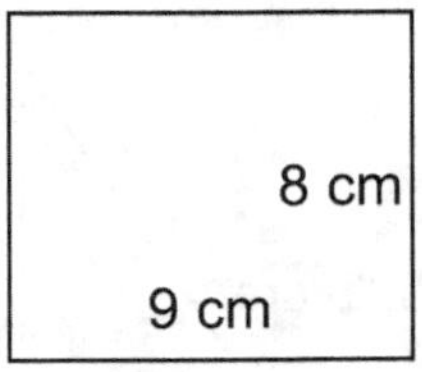 8 cm, 9 cm

Find the area.

1.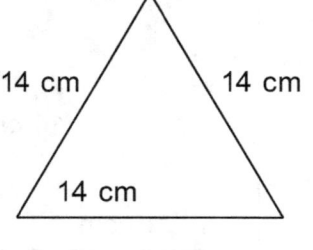
....

2.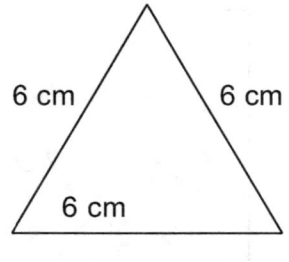
....

3.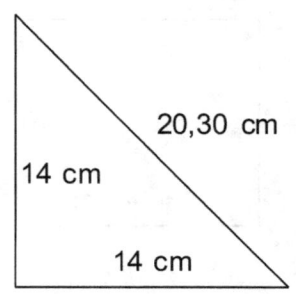
....

4.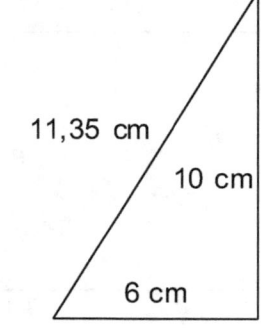
....

5.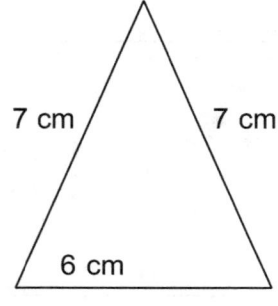
....

6.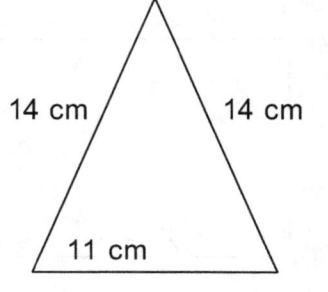
....

7.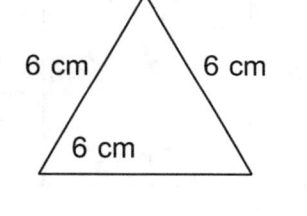
....

8.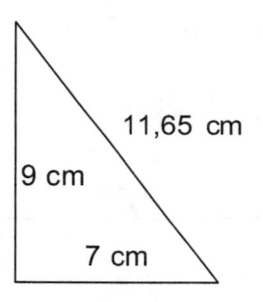
....

9.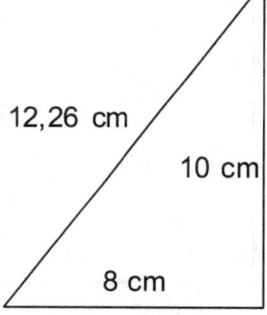
....

10.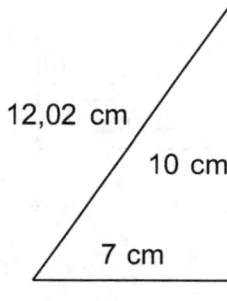
....

11.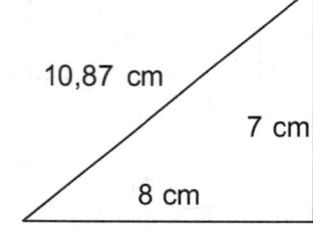
....

12.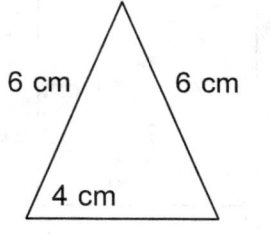
....

Find the area.

1.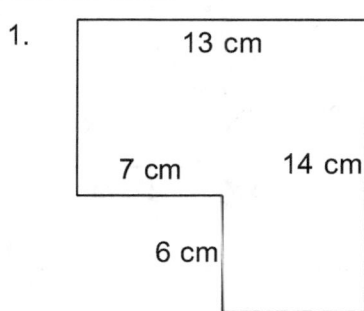
2.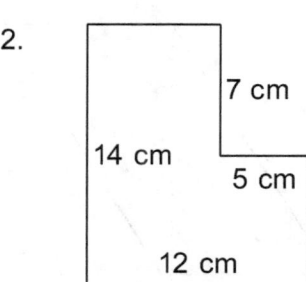
3.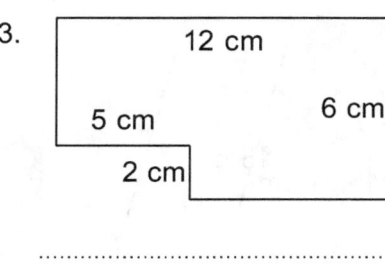

4.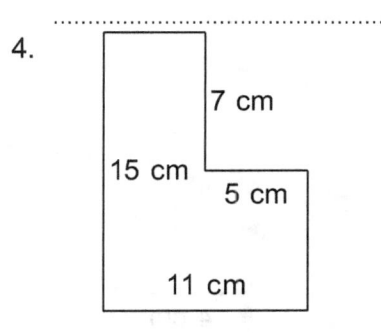
5.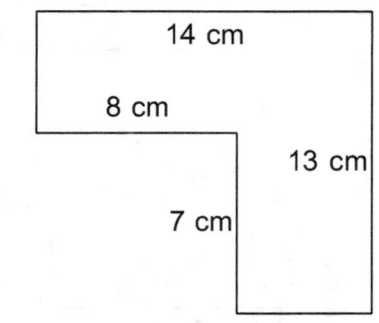
6.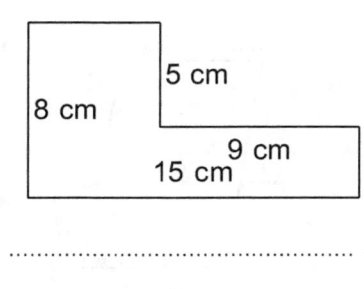

7.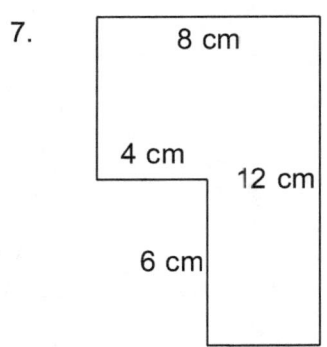
8.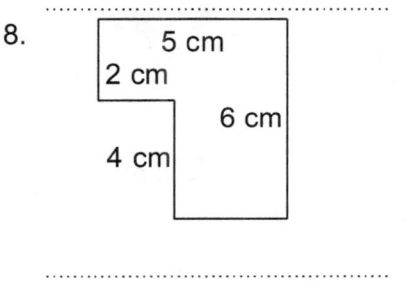
9.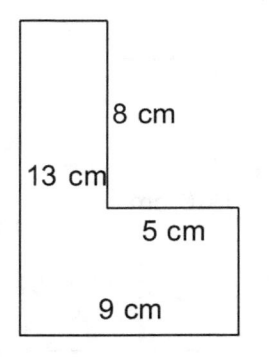

10.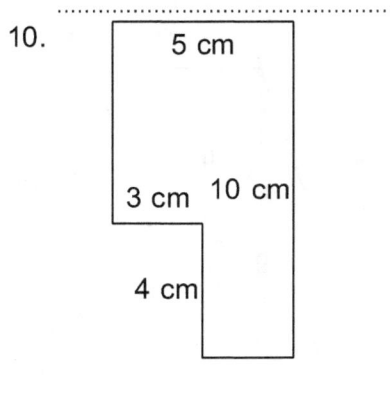
11.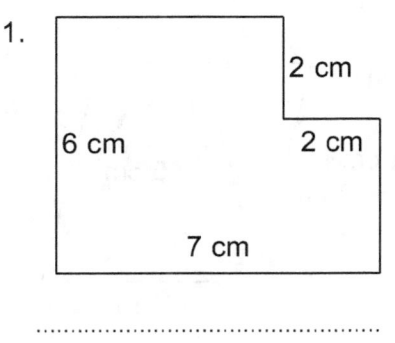
12.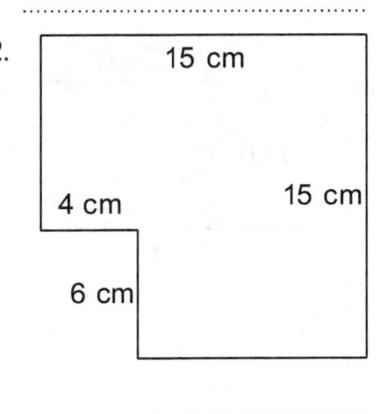

© KingSchool Edition

Find the area.

1.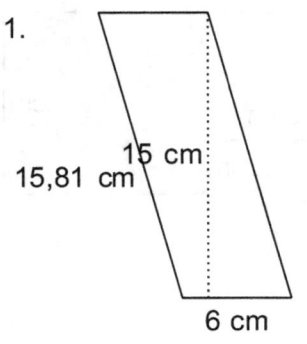

2.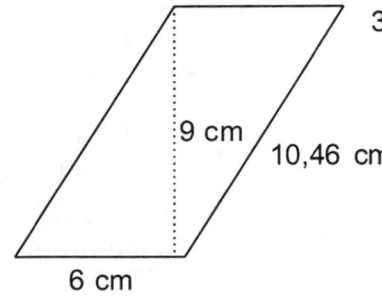

3.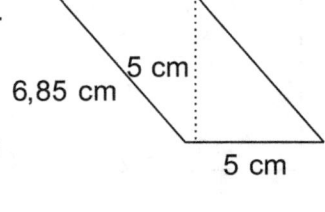

4.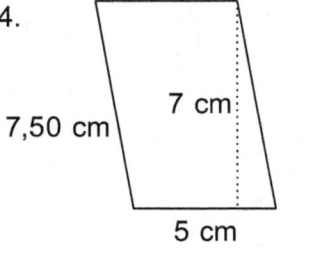

5.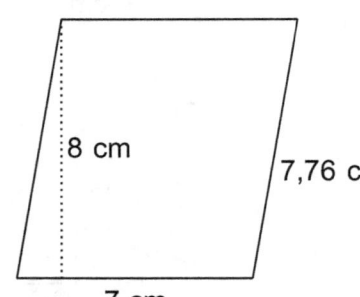

6.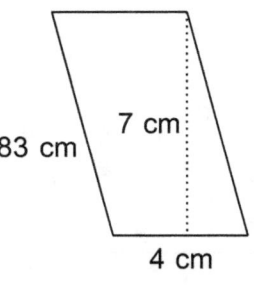

7.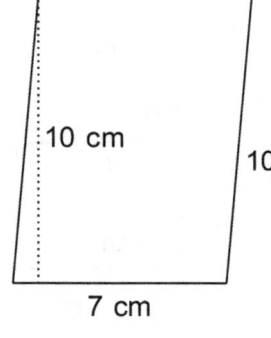

8.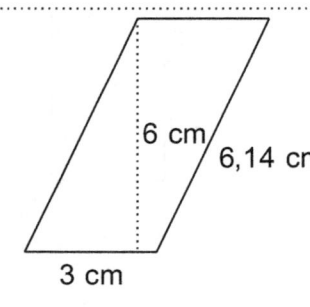

9.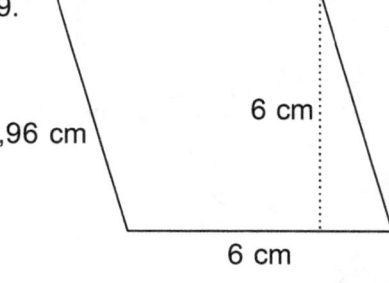

10.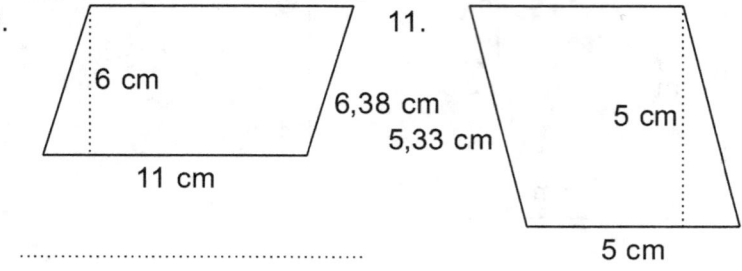

11.

12.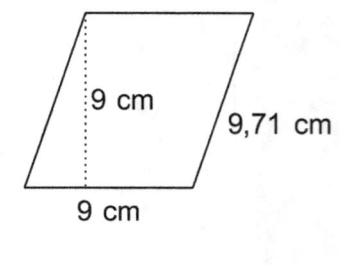

Find the volume.

1.

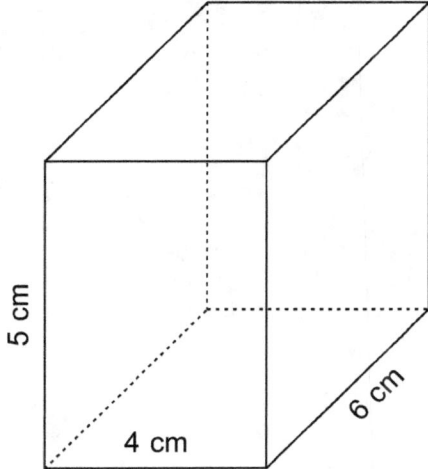

2.

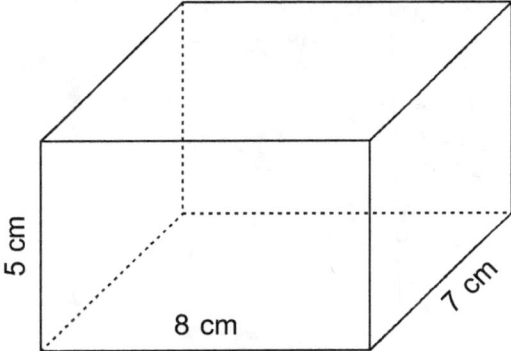

3.

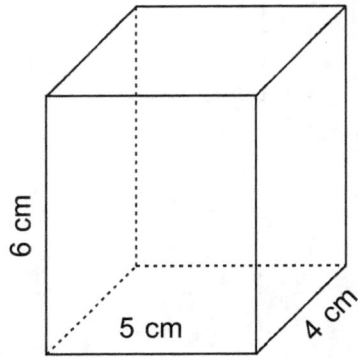

Find the volume.

1.

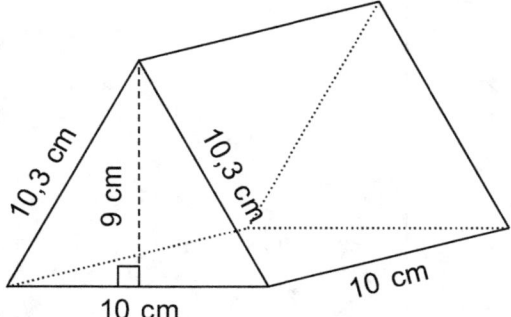

2.

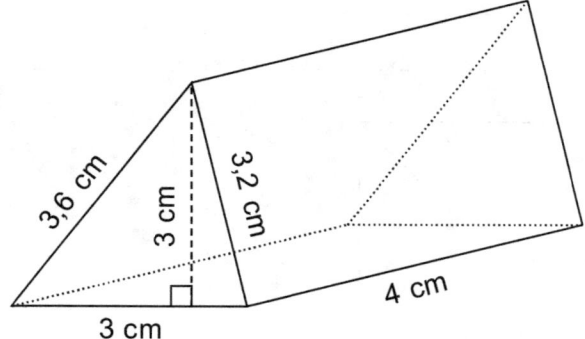

3.

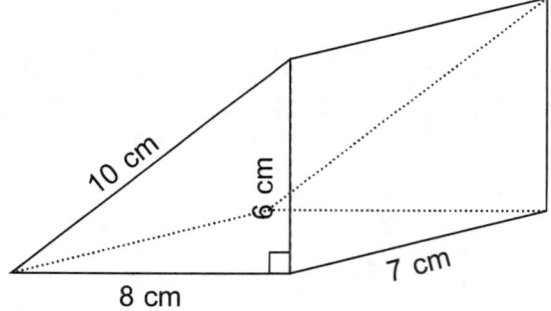

Find the volume.

1.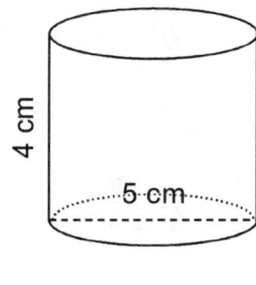
....

2.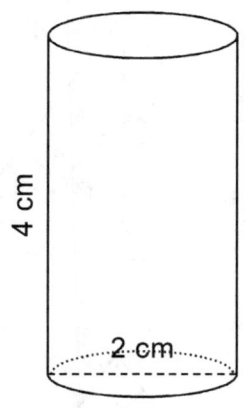
....

3.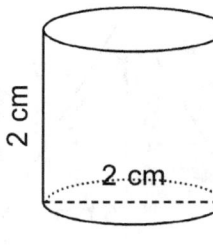
....

4.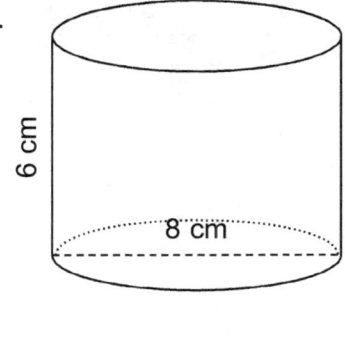
....

5.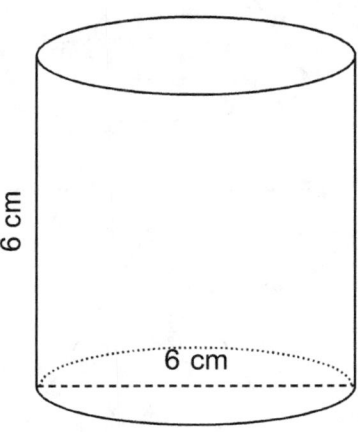
....

6.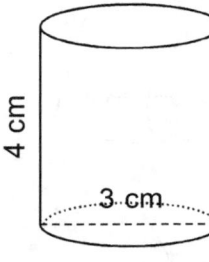
....

7.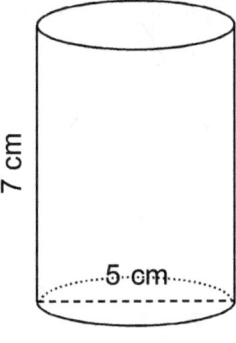
....

8.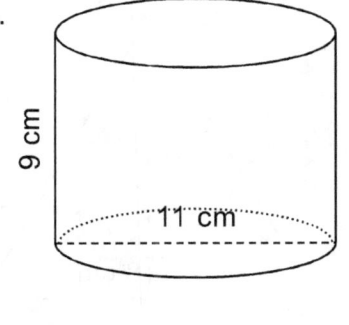
....

9.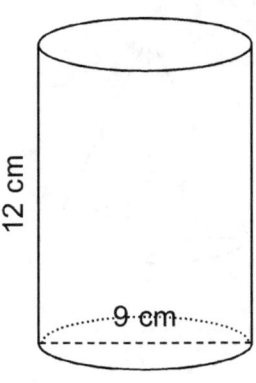
....

Find the volume.

1.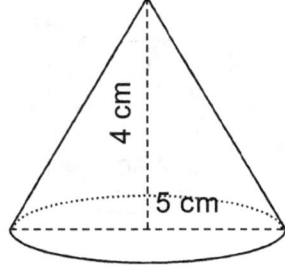

2.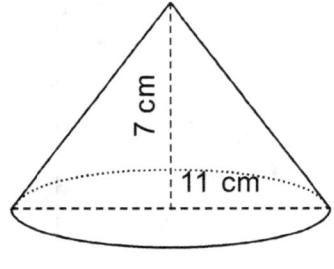

3.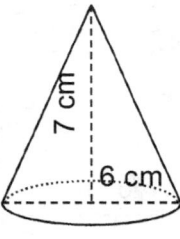

4.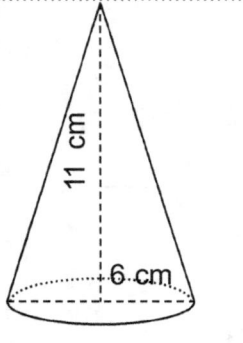

5.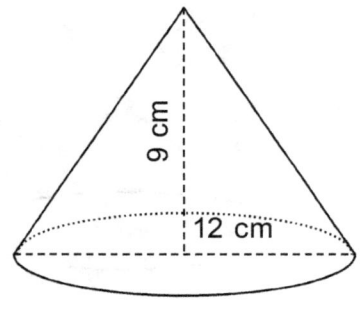

6.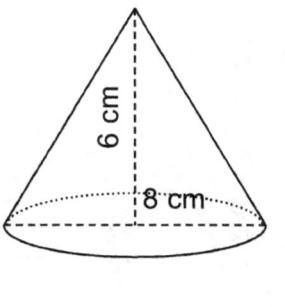

7.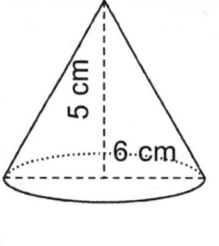

8.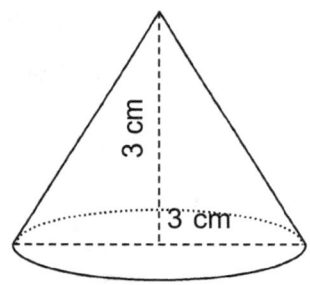

9.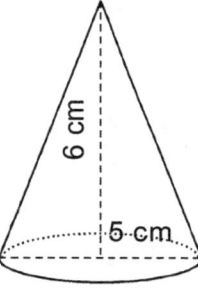

10.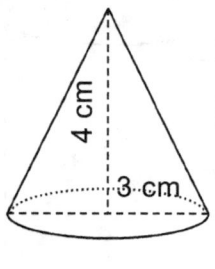

11.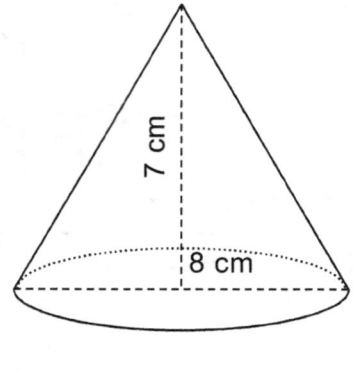

12.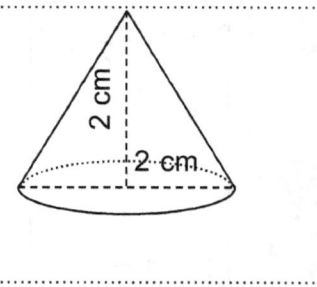

Find the volume.

1.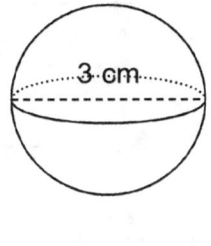
....

2.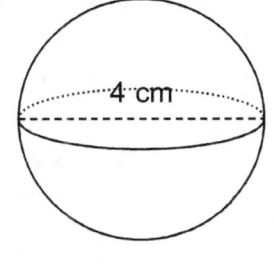
....

3.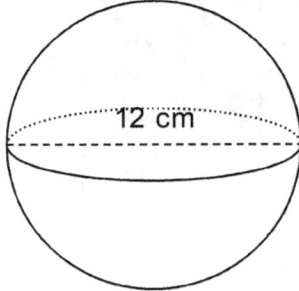
....

4.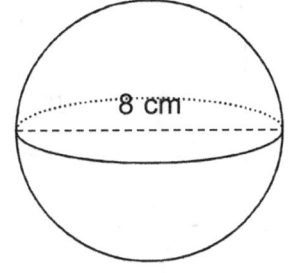
....

5.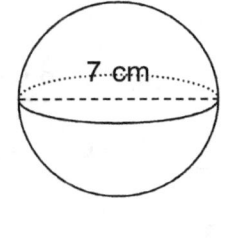
....

6.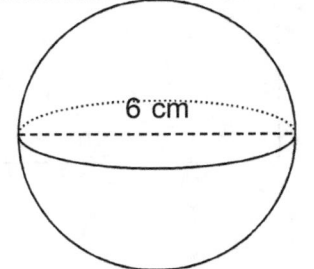
....

7.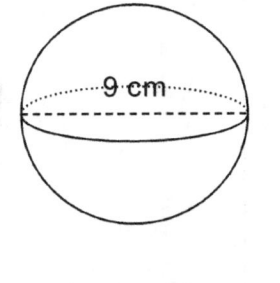
....

8.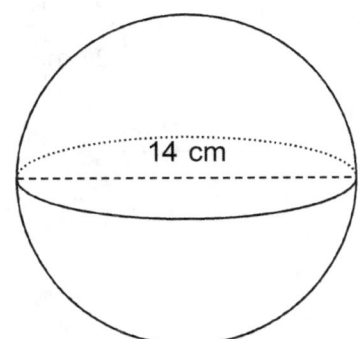
....

9.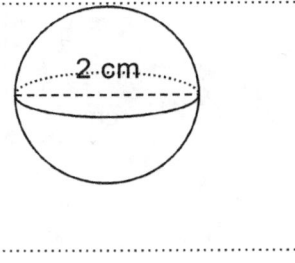
....

10.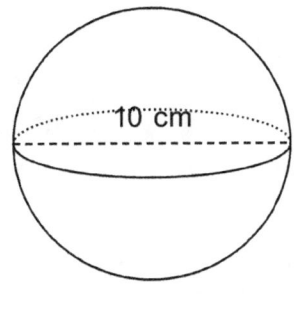
....

11.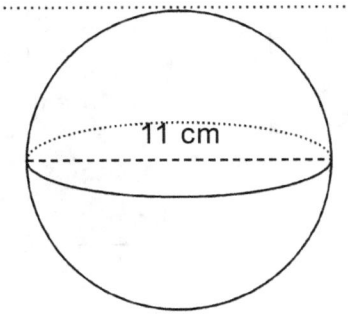
....

12.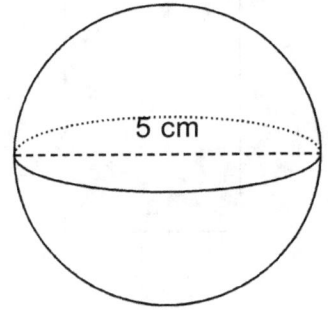
....

Find the perimeter.

1. 7 cm, 10 cm
   ....

2. 5 cm, 6 cm
   ....

3. 8 cm, 7 cm
   ....

4. 9 cm, 16 cm
   ....

5. 8 cm, 10 cm
   ....

6. 9 cm, 8 cm
   ....

7. 7 cm, 7 cm
   ....

8. 9 cm, 10 cm
   ....

9. 8 cm, 6 cm
   ....

10. 11 cm, 7 cm
    ....

11. 13 cm, 11 cm
    ....

12. 10 cm, 8 cm
    ....

Find the perimeter.

1.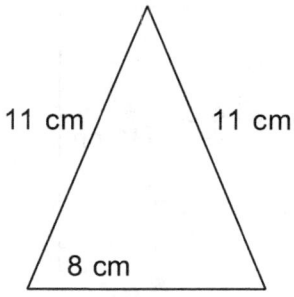

2.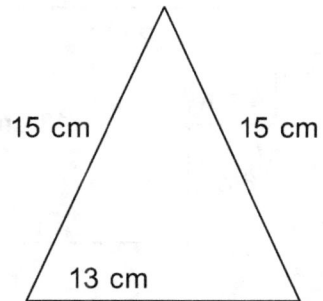

3.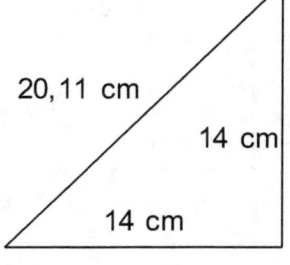

4.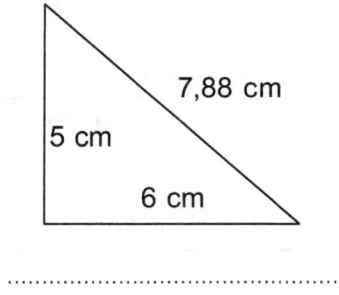

5.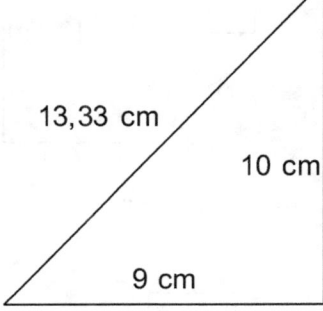

6.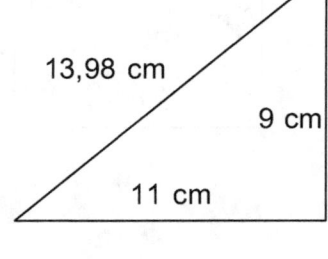

7.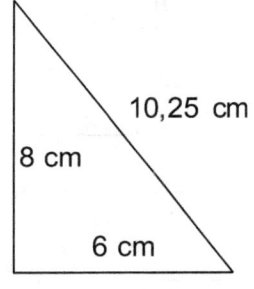

8.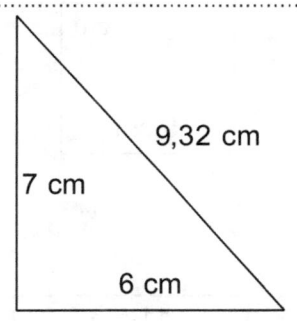

9.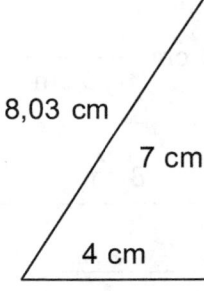

10.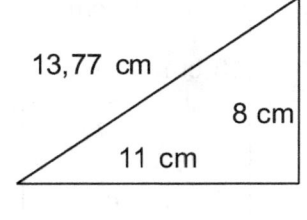

11.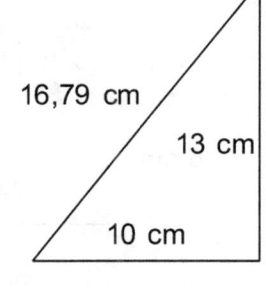

12.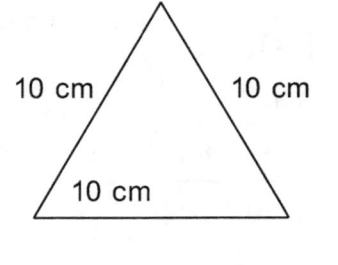

Find the perimeter.

1.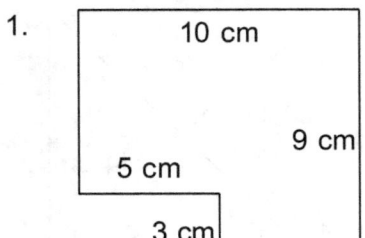

2.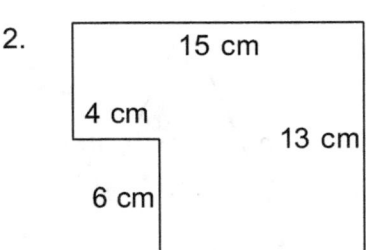

3.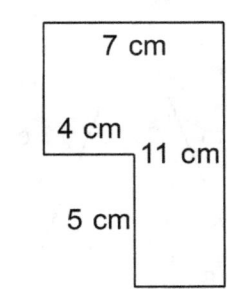

4.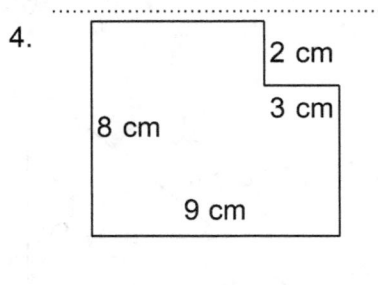

5.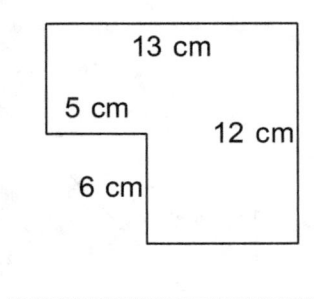

6.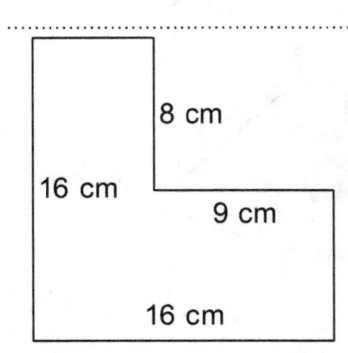

7.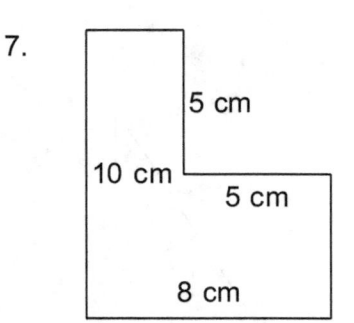

8.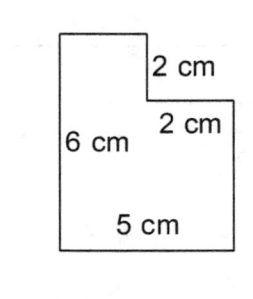

9.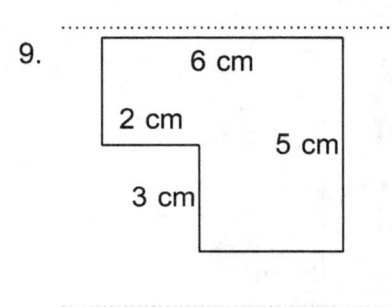

10.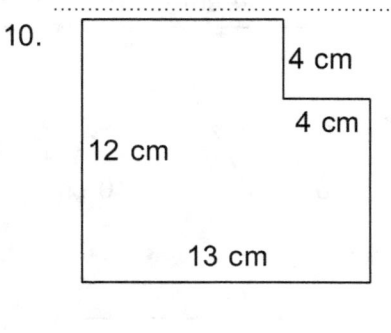

11.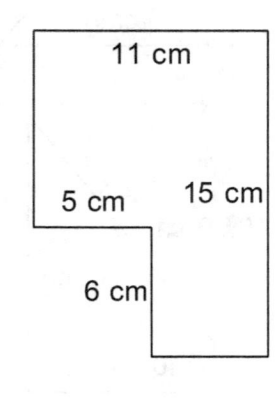

12.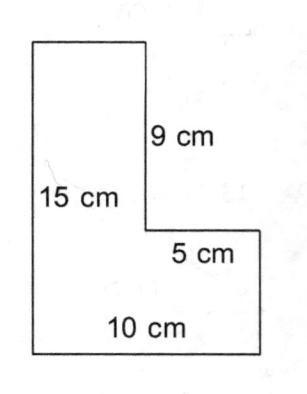

Find the perimeter.

1.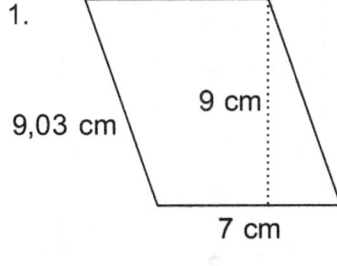

2.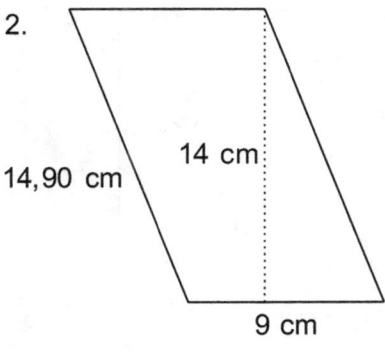

3.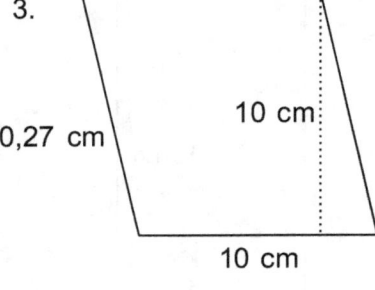

4.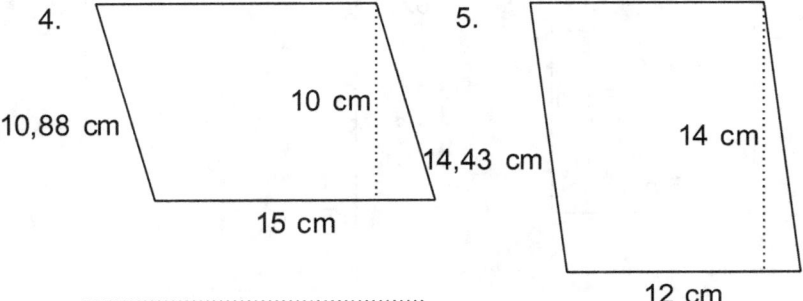

5.

6.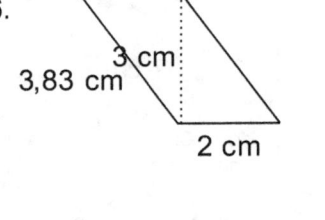

7.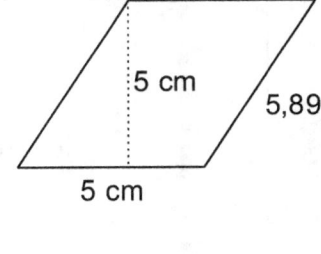

8.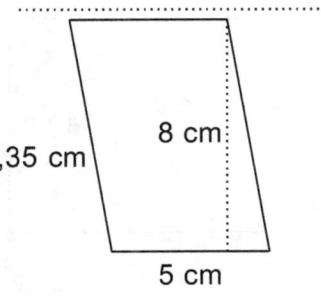

9.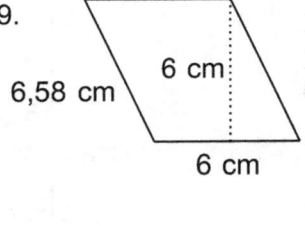

10.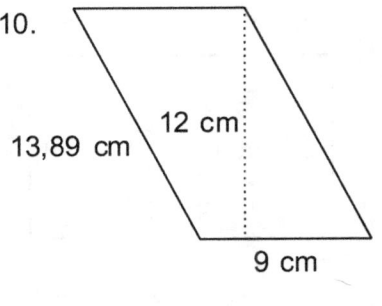

11.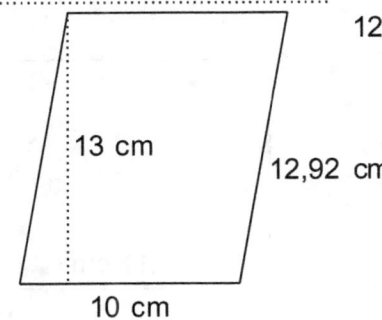

12.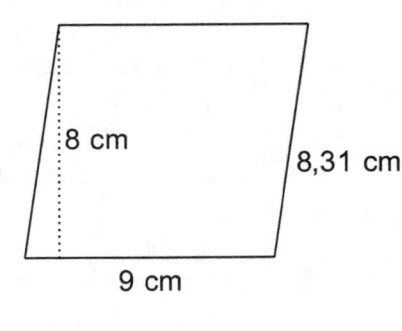

Find the area.

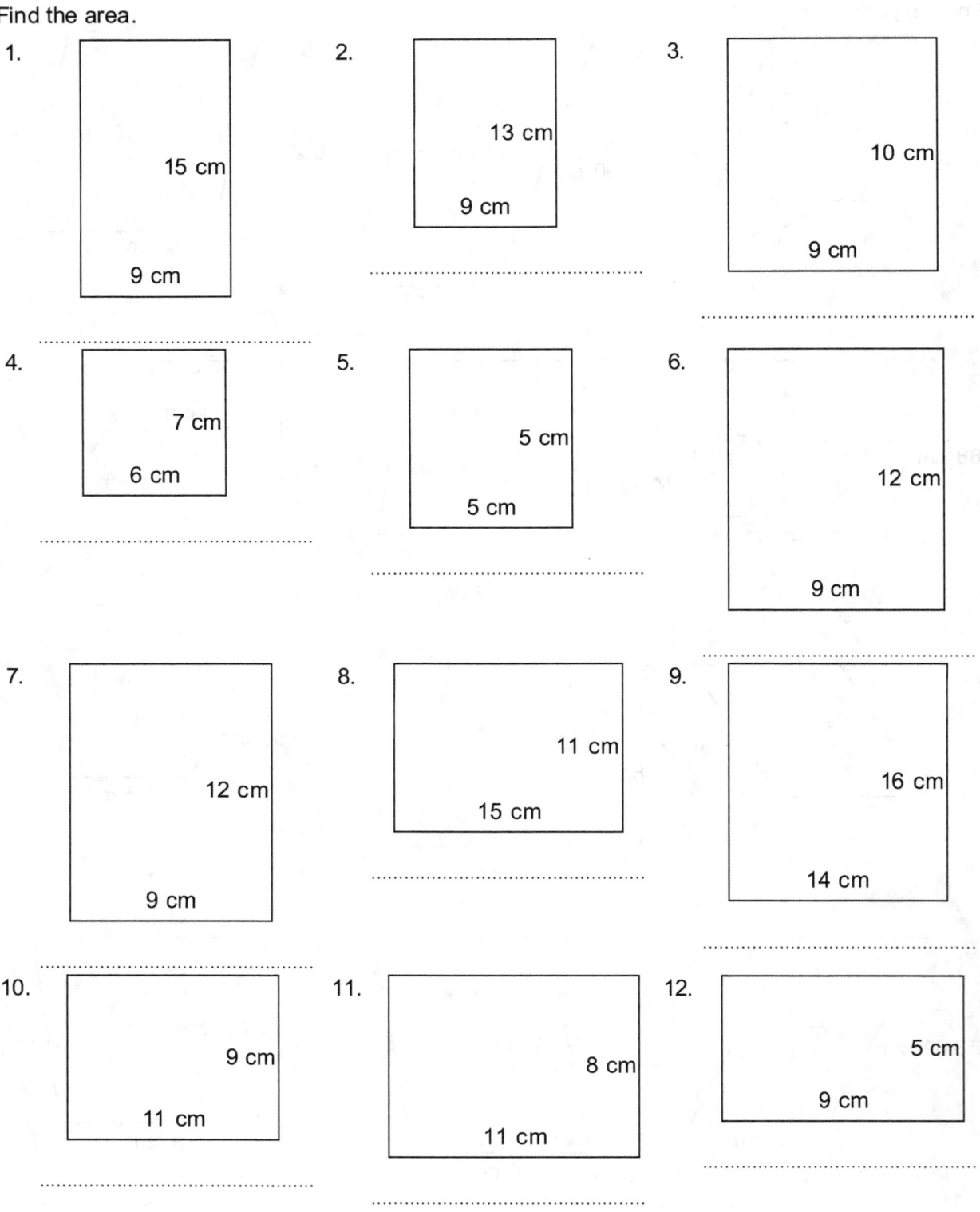

Find the area.

1.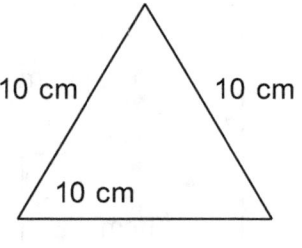
....

2.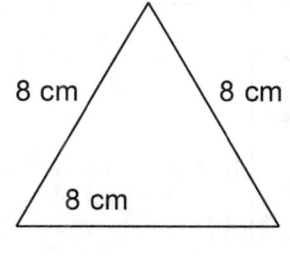
....

3.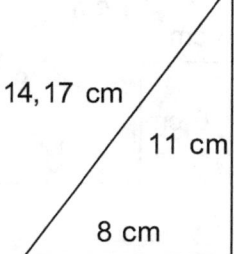
....

4.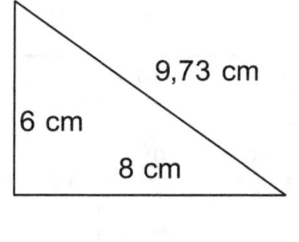
....

5.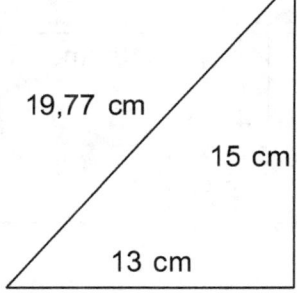
....

6.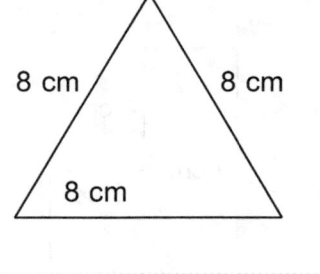
....

7.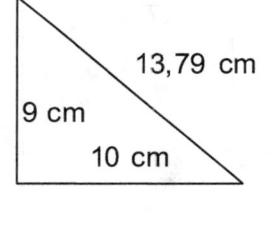
....

8.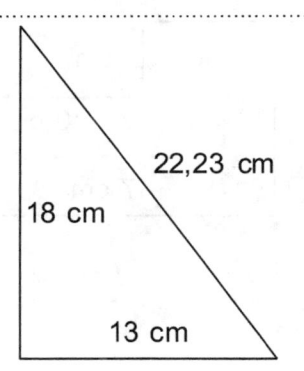
....

9.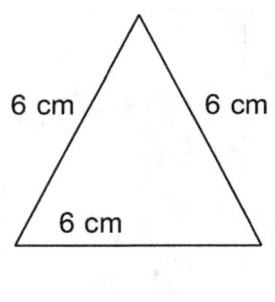
....

10.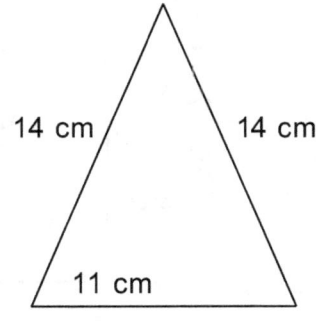
....

11.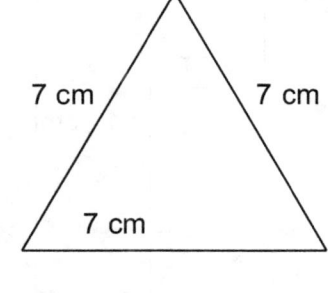
....

12.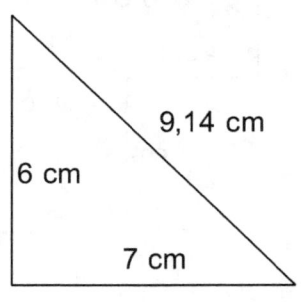
....

Find the area.

1.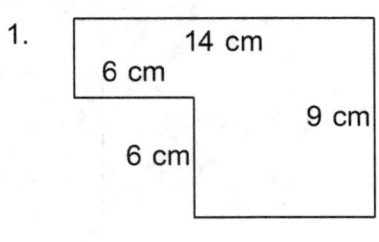

2.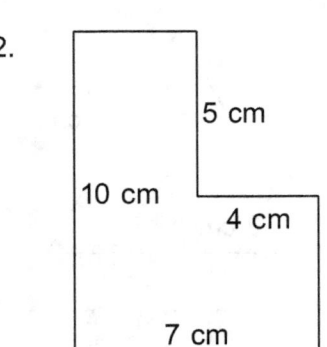

3.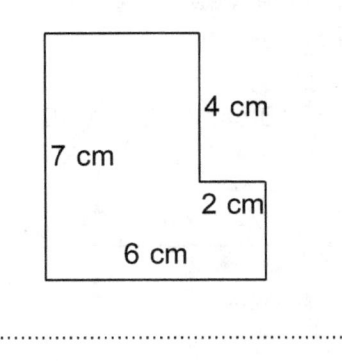

4.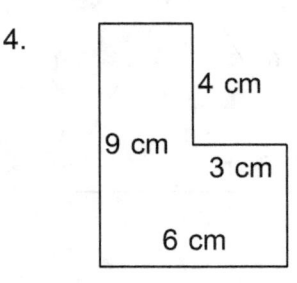

5.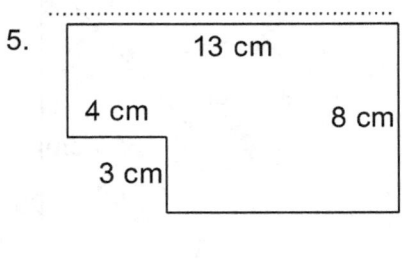

6.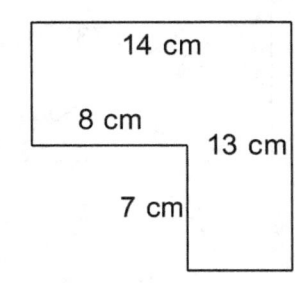

7.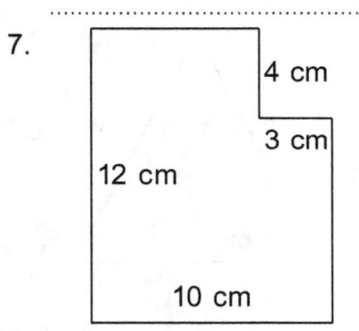

8.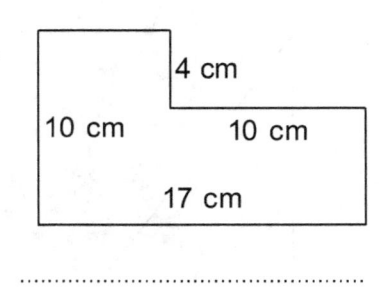

9.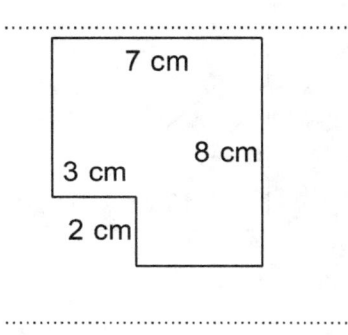

10.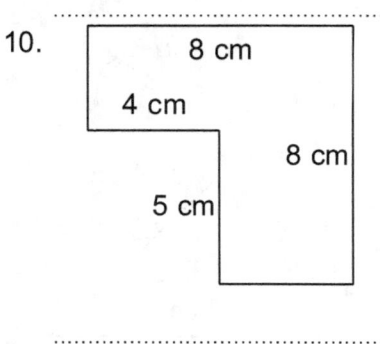

11.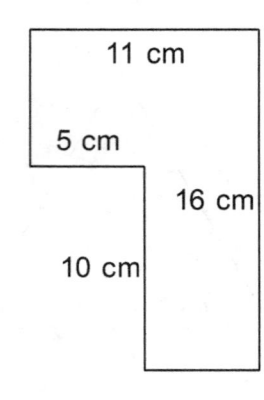

12.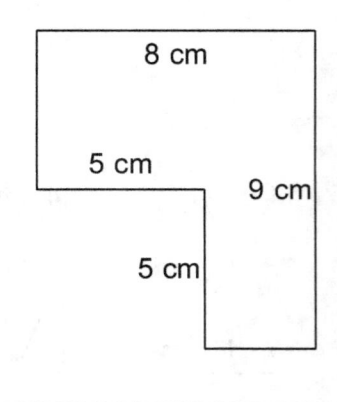

Find the area.

1.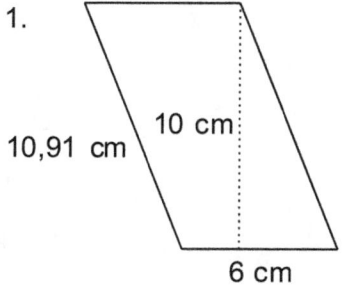

2.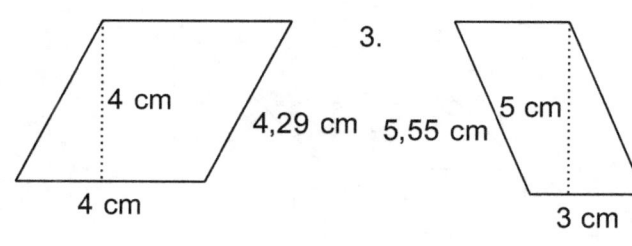

3.

4.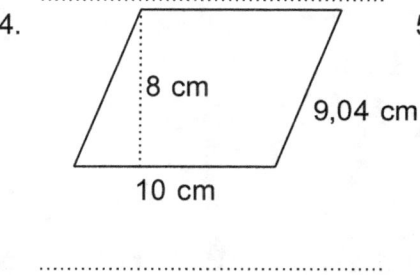

5.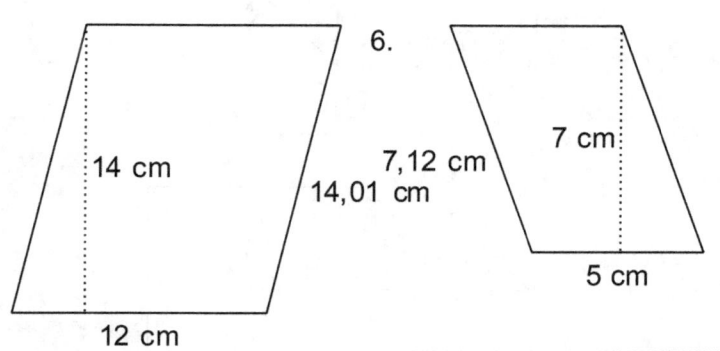

6.

7.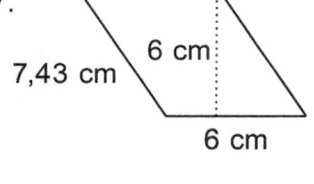

8.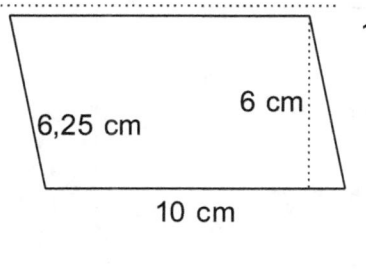

9.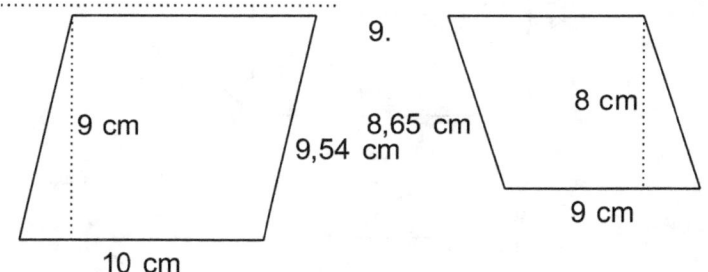

10.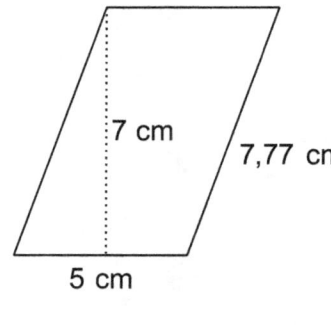

11.

12.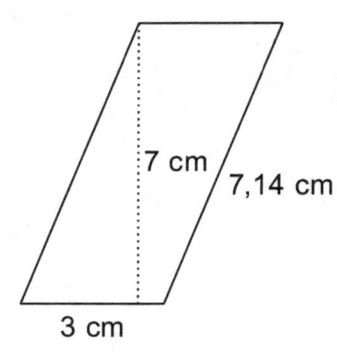

Find the volume.

1.
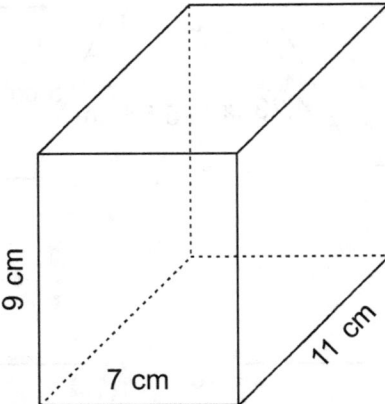

2.
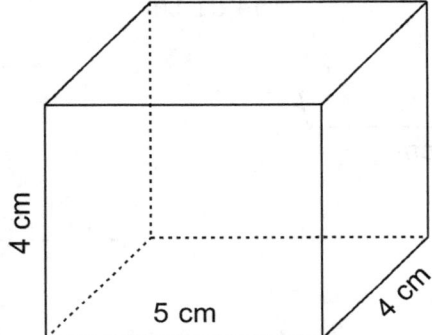

3.
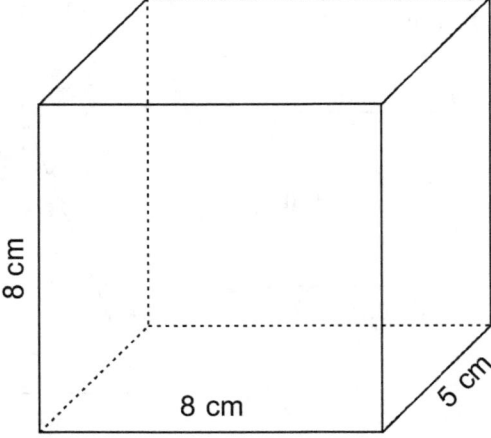

Find the volume.

1.

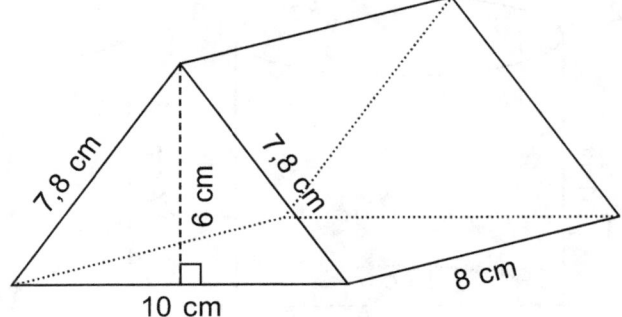

2.

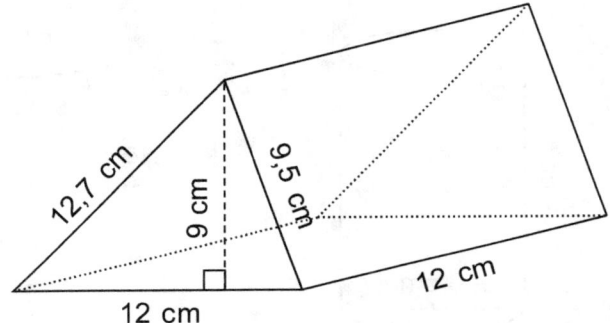

3.

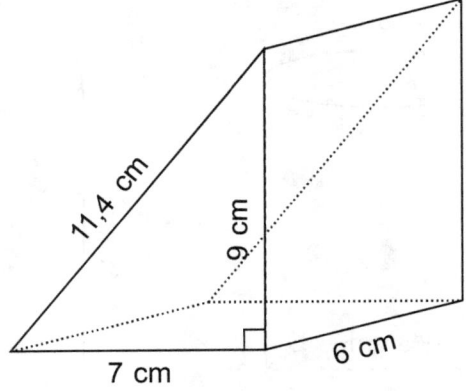

Find the volume.

1.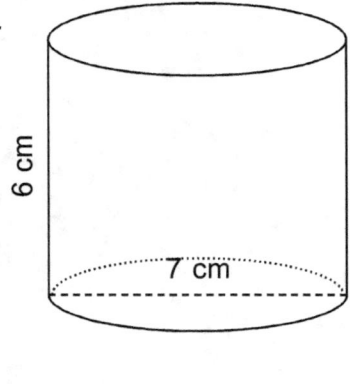

2.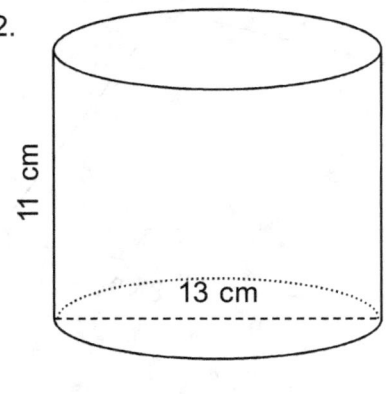

3.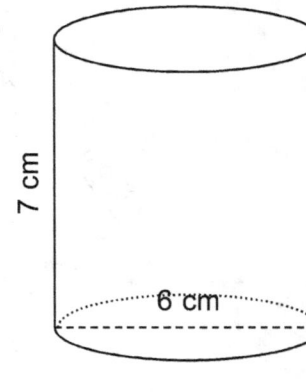

4.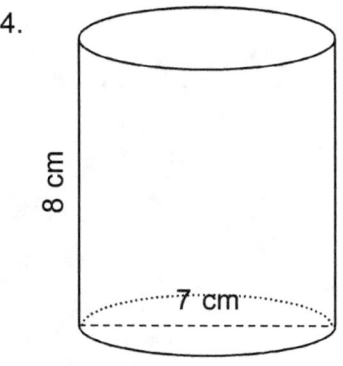

5.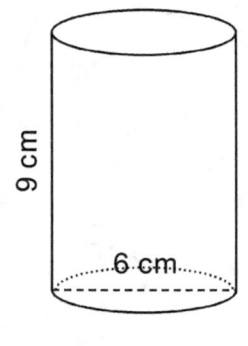

6.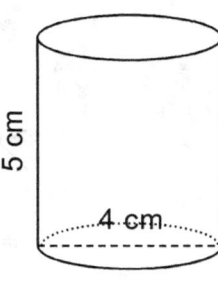

7.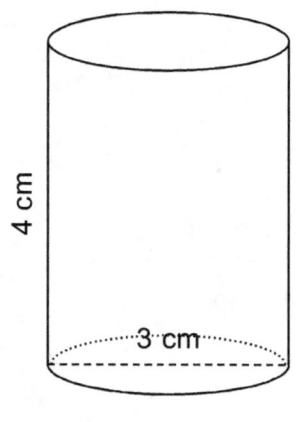

8.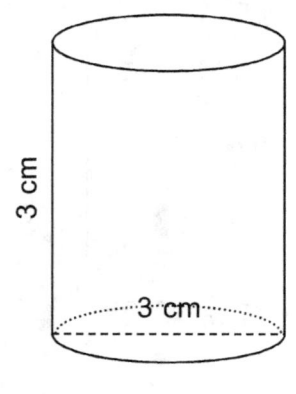

9.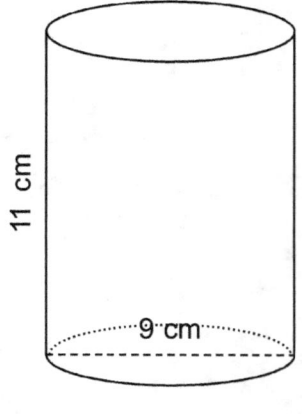

Find the volume.

1.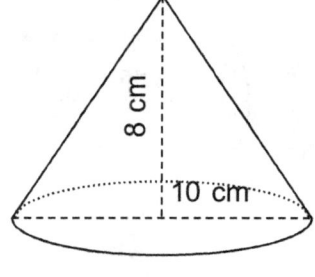
....

2.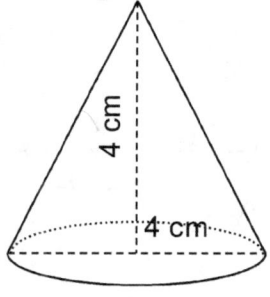
....

3.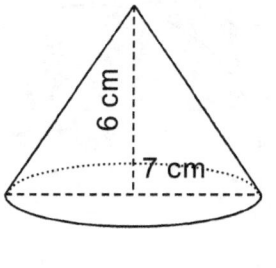
....

4.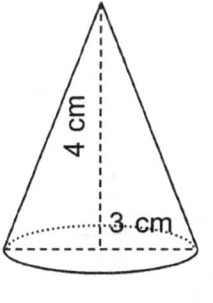
....

5.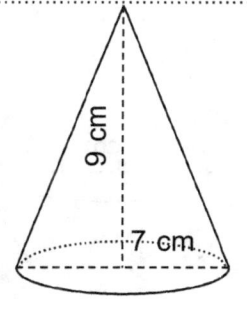
....

6.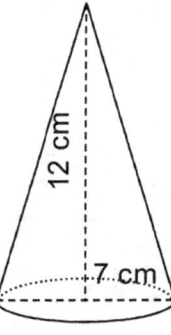
....

7.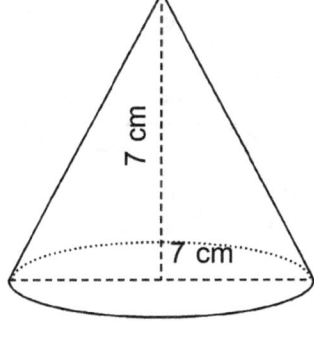
....

8.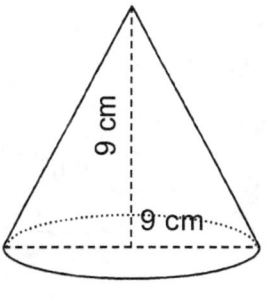
....

9.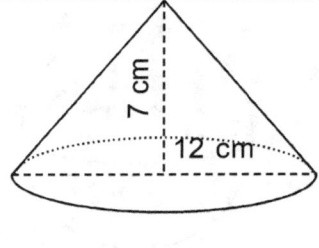
....

10.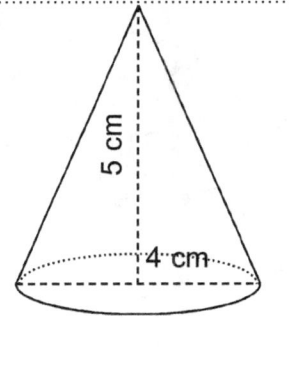
....

11.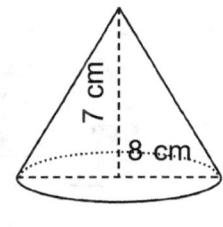
....

12.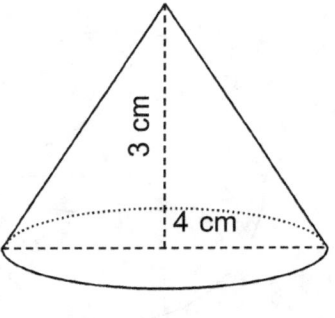
....

Find the volume.

1.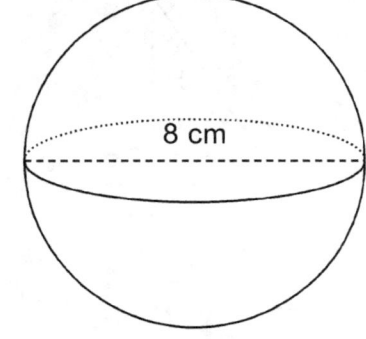
2.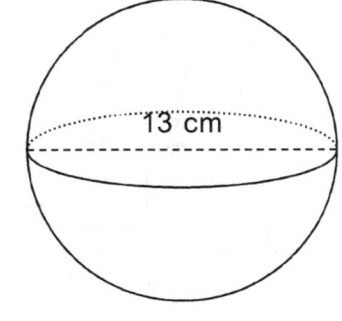
3.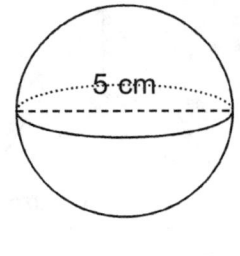

4.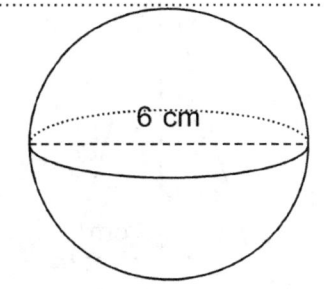
5.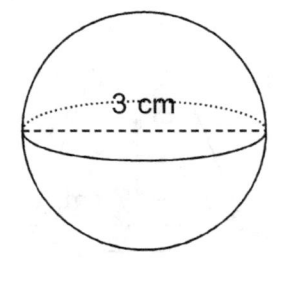
6.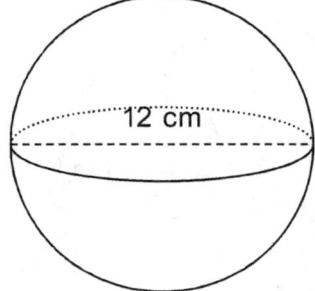

7.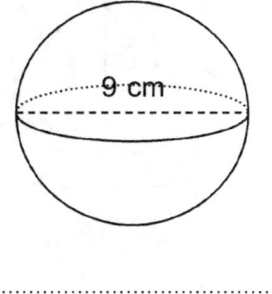
8.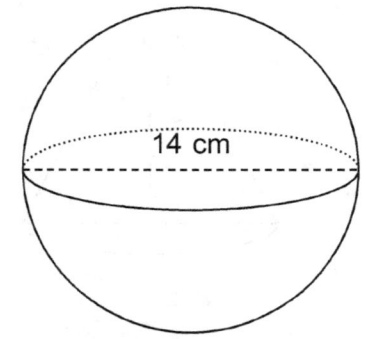
9.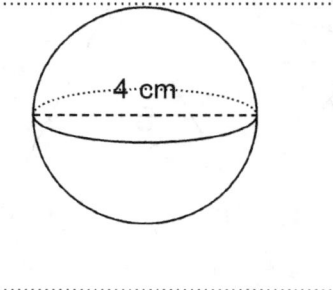

10.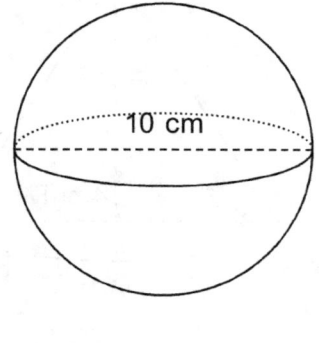
11.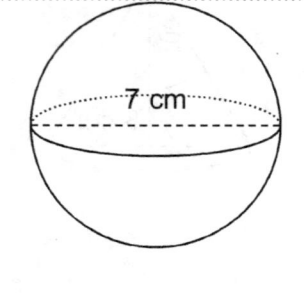
12.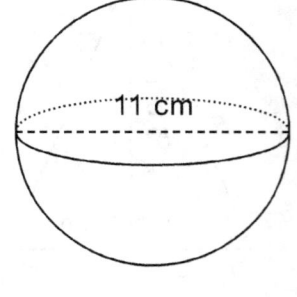

© KingSchool Edition

www.ingramcontent.com/pod-product-compliance
Lightning Source LLC
Chambersburg PA
CBHW080620220526

45466CB00010B/3403